SCIENCE QUIZ WHIZ

WRITTEN BY
Linda Schwartz

ILLUSTRATED BY
Kelly Kennedy

The
Learning
Works

Illustrator: Kelly Kennedy

Book design: Studio E Books, Santa Barbara, CA

Cover illustrator: Rick Grayson

Cover designer: Barbara Peterson

Project director: Linda Schwartz

© 2003 The Learning Works, Inc.

The purchase of this book entitles the individual teacher to reproduce copies for use in the classroom.

Reproduction of any part for an entire school or school system or for commercial use is strictly prohibited.

No form of this work may be reproduced, transmitted, or recorded without written permission from the publisher.

CONTENTS

The question cards in *Science Quiz Whiz* are grouped into 10 categories of 36 question cards each. In each section you will find 6 folios, each with 6 cards. The 10 categories can be identified by their borders, as shown below:

WAYS TO USE *Science Quiz Whiz* IN CLASS

There are numerous ways to use *Science Quiz Whiz* in class. To begin with, you can open the book to any page and ask a few questions to start your morning or to begin a science lesson, or to fill those last minutes before lunch, recess, or the end of the day. If you have more time, try one of the following creative ideas.

Science Quiz Whiz Game

Start by removing the pages from this book and cutting the question cards apart. (If you prefer to keep the book intact, simply photocopy the question cards from the section or sections you wish to use.) For added durability, laminate the pages before you cut the cards apart. To help you identify topic categories, a different border has been used for each of the ten sections of the book.

Make a bulletin board display using the headers provided on pages 6 and 7. Divide the headers with colored yarn as shown. Select five categories at a time (or more, if you prefer), and pin five question cards from the five categories under each heading. Attach an unlined index card with a dollar value written on it over each question card. The more difficult questions should be worth more money and should be placed farther down on the quiz board.

Once the quiz board is set up, it can be used over and over by simply changing the category headers and/or replacing question cards with new ones.

Students can get together and decide on rules of the game. Encourage them to think of other topics that relate to science and to add their own question cards to the game board. A moderator can be selected, and someone can be assigned to check to see if the questions have been answered correctly by simply looking on the reverse side of the question cards on the board.

Select a scorekeeper to keep track of money earned. You can also use play money as Quiz Whiz bucks to award players. Play money can be found at many school supply and toy stores.

Triple Tic-Tac-Toe

This is a great game for students to play at a science center when they've completed their class assignments. Reproduce copies of the Triple Tic-Tac-Toe game board on page 8, and place them at the center along with the question cards. Students play with partners, taking turns drawing questions from the pile and answering them. If a player answers a question correctly, he or she marks an X or an O in pencil on any one of the three Tic-Tac-Toe grids. Students must play defensively, trying to block their opponents from getting three correct answers in a row in any direction while attempting to score Tic-Tac-Toe themselves.

Science Quiz Whiz Bee

Use the questions for a Science Quiz Whiz Bee, organized like a spelling bee. Contestants are eliminated as they miss questions asked by a student moderator. Have class champs challenge each other, or organize a school-wide Science Quiz Whiz Bee.

Radio or Television Game Show

Use Science Quiz Whiz questions to organize a classroom quiz show in a radio or television format. Select one student to be the "host" and four to six students to be "contestants." Create your own rules: correct answers can earn Quiz Whiz Bucks; incorrect answers might lose Bucks. The end of the game can come after a predetermined period of time, or after a certain number of questions have been asked and answered.

Science Quiz Whiz Question of the Day

Select some of the most difficult questions and use them as research challenges for a class assignment each day. Students can work alone or with partners to answer the question of the day. This is also a good way to get students to seek information on the Internet. Award Quiz Whiz Bucks to the first student(s) to find the correct answer. Quiz Whiz Bucks can be redeemed for awards at the end of the week. A variation would be to select a difficult question each day for homework or for an extra-credit challenge.

Plants

Insects

Amphibians & Reptiles
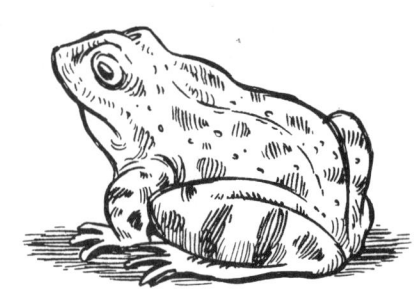

Birds & Fishes

Mammals

The Human Body
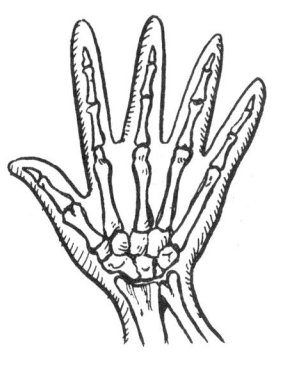

Inventions & Discoveries

Electricity & Energy

Weather & Water

The Universe

Triple Tic-Tac-Toe Game Board

SCIENCE
Quiz Whiz

What are cone-bearing plants called?

SCIENCE
Quiz Whiz

What is the name of the green material in plants that gives leaves their color and enables them to make food?

SCIENCE
Quiz Whiz

What kind of tree produces acorns?

SCIENCE
Quiz Whiz

What name is given to trees that stay green throughout the year?

SCIENCE
Quiz Whiz

What name is given to a person who studies plants?

SCIENCE
Quiz Whiz

What is the process called by which green plants convert sunlight to energy?

SCIENCE
Quiz Whiz

What three things are needed by a plant for photosynthesis to occur?

SCIENCE
Quiz Whiz

Which of the following foods does *not* come from the root of the plant: cucumbers, beets, carrots, or sweet potatoes?

SCIENCE
Quiz Whiz

Which of the following is *not* a fruit: banana, asparagus, avocado, or apple?

SCIENCE
Quiz Whiz

What are the leaves of pine trees called?

SCIENCE
Quiz Whiz

What term describes leaves that are composed of two or more leaflets on a common stalk?

SCIENCE
Quiz Whiz

Which of the following fruits does *not* grow on a vine: cranberry, tomato, watermelon, or lemon?

SCIENCE
Quiz Whiz
cucumbers

SCIENCE
Quiz Whiz
air, water, and sunlight

SCIENCE
Quiz Whiz
needles

SCIENCE
Quiz Whiz
asparagus

SCIENCE
Quiz Whiz
lemon

SCIENCE
Quiz Whiz
compound leaves

SCIENCE
Quiz Whiz

What part of a flower holds pollen?

SCIENCE
Quiz Whiz

What is the process called by which plants give off water?

SCIENCE
Quiz Whiz

What is another name for the leaves of ferns?

SCIENCE
Quiz Whiz

Which of the following fruits does *not* grow on trees: grape, apple, avocado, or cherry?

SCIENCE
Quiz Whiz

What is the center part of a flower called that is often sticky and makes seeds?

SCIENCE
Quiz Whiz

Which of the following foods does *not* come from a plant's leaves: lettuce, cauliflower, spinach, or cabbage?

SCIENCE Quiz Whiz

transpiration

SCIENCE Quiz Whiz

stamen

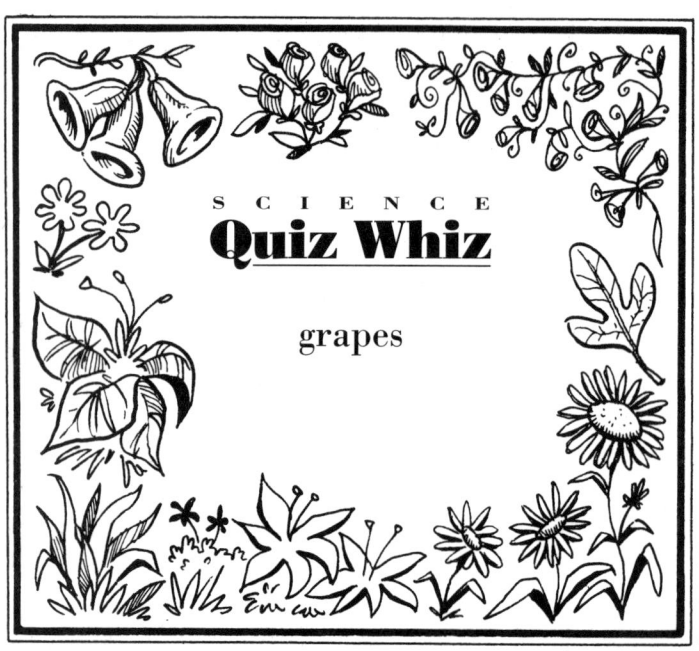

SCIENCE Quiz Whiz

grapes

SCIENCE Quiz Whiz

fronds

SCIENCE Quiz Whiz

cauliflower

SCIENCE Quiz Whiz

pistil

SCIENCE Quiz Whiz

What is another name for the central vein of a leaf, which runs from the base to the tip?

SCIENCE Quiz Whiz

Which of the following foods is *not* a seed: corn, plum, pea, or lima bean?

SCIENCE Quiz Whiz

What is the name of a young branch or root taken from an old plant for propagation?

SCIENCE Quiz Whiz

What part of a plant anchors it in the ground and derives nutrients from the soil?

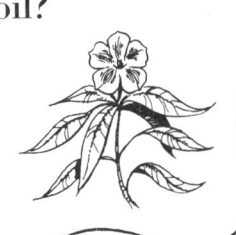

SCIENCE Quiz Whiz

What part of a plant makes most of the food that it needs to live and grow?

SCIENCE Quiz Whiz

What reproductive process do plants accomplish with the help of bees and butterflies?

SCIENCE
Quiz Whiz
plum

SCIENCE
Quiz Whiz
midrib

SCIENCE
Quiz Whiz
roots

SCIENCE
Quiz Whiz
a cutting (or slip)

SCIENCE
Quiz Whiz
pollination

SCIENCE
Quiz Whiz
leaf

SCIENCE Quiz Whiz

What is the name of the special leaves that cover and protect a flower bud while it is growing?

SCIENCE Quiz Whiz

Which of the following foods does *not* come from the stalks or stems of plants: rhubarb, asparagus, celery, or turnip?

SCIENCE Quiz Whiz

What part of a plant holds up the leaves and flowers and carries water and minerals from the roots to the leaves?

SCIENCE Quiz Whiz

Which of the following is *not* an insect-eating plant: Venus's flytrap, dandelion, sundew, or pitcher plant?

SCIENCE Quiz Whiz

Which of the following plants is *not* a type of cactus: prickly pear, saguaro, barrel, or hickory?

SCIENCE Quiz Whiz

What is the stiff, woody stem of a tree called?

SCIENCE
Quiz Whiz
turnip

SCIENCE
Quiz Whiz
sepals

SCIENCE
Quiz Whiz
dandelion

SCIENCE
Quiz Whiz
stem

SCIENCE
Quiz Whiz
trunk

SCIENCE
Quiz Whiz
hickory

SCIENCE Quiz Whiz

Which of the following fruits does *not* grow on a bush: raspberry, peach, strawberry, or blueberry?

SCIENCE Quiz Whiz

Which of the following plants is *not* a grain: wheat, corn, rice, or onion?

SCIENCE Quiz Whiz

What kind of tree found in California is among the largest living things on earth?

SCIENCE Quiz Whiz

What is the thick, fat part of an underground stem called, such as a potato?

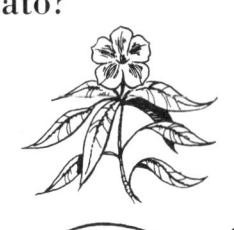

SCIENCE Quiz Whiz

What fruit has seeds on the outside: watermelon, strawberry, lemon, or orange?

SCIENCE Quiz Whiz

Which of the following is a member of the rose family: peach, artichoke, apple, or orange?

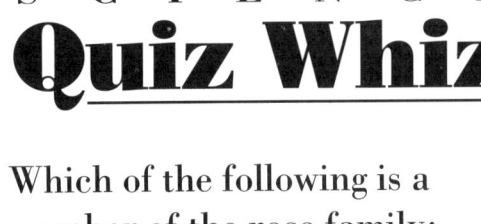

SCIENCE Quiz Whiz

The body of an insect consists of how many segments?

SCIENCE Quiz Whiz

Which of the following is *not* an insect: spider, cricket, mosquito, or flea?

SCIENCE Quiz Whiz

How many legs do insects have?

SCIENCE Quiz Whiz

Which are *not* arachnids: spiders, ticks, scorpions, or slugs?

SCIENCE Quiz Whiz

Which of the following is *not* considered a social insect: termite, firefly, bee, or ant?

SCIENCE Quiz Whiz

What is another name for an insect's feelers?

SCIENCE Quiz Whiz

Which of the following is *not* an order, or group, of insects: beetles, black widows, termites, or ants?

SCIENCE Quiz Whiz

How many wings do butterflies have?

SCIENCE Quiz Whiz

What is the name of the large groups in which social insects such as ants and bees live?

SCIENCE Quiz Whiz

What does the word "metamorphosis" mean?

SCIENCE Quiz Whiz

What is the main job of a queen bee or termite?

SCIENCE Quiz Whiz

What is the name of the wormlike animal that hatches from an insect's egg?

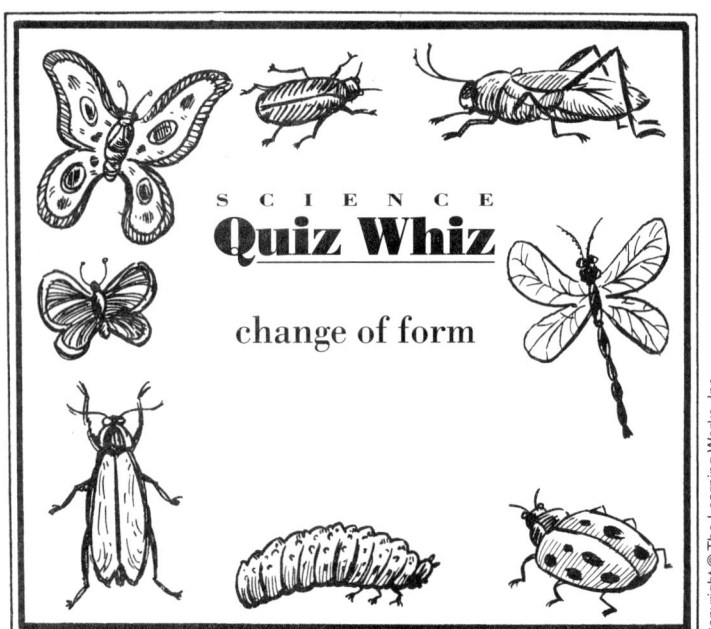

SCIENCE
Quiz Whiz

What is the larva of a butterfly or moth called?

SCIENCE
Quiz Whiz

In complete metamorphosis, what stage follows the larval stage?

SCIENCE
Quiz Whiz

Which of the following insects does *not* undergo complete metamorphosis: butterfly, moth, grasshopper, or ant?

SCIENCE
Quiz Whiz

The honeycomb built by bees is made of what substance?

SCIENCE
Quiz Whiz

How many pairs of wings do flies have?

SCIENCE
Quiz Whiz

What is the fastest-flying insect?

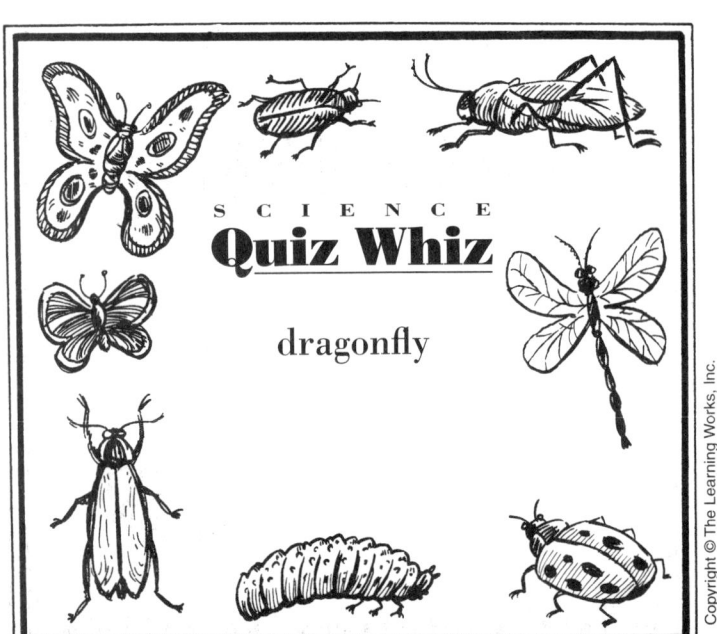
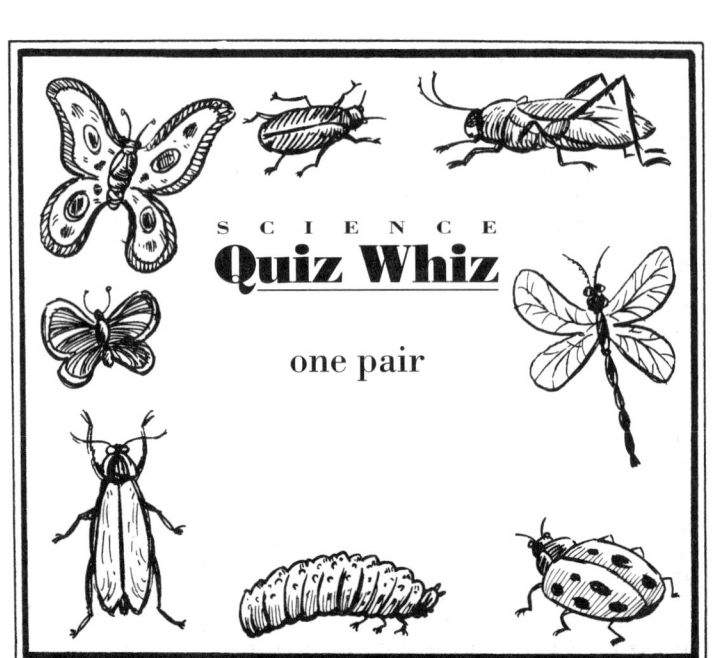

SCIENCE
Quiz Whiz

What tiny holes on an insect's body help it breathe?

SCIENCE
Quiz Whiz

Which is *not* a color of insects' blood: yellow, red, or green?

SCIENCE
Quiz Whiz

On what part of the body are the ears of long-horned grasshoppers and crickets located?

SCIENCE
Quiz Whiz

Which of the following is *not* a type of beetle: mosquito, firefly, weevil, or ladybug?

SCIENCE
Quiz Whiz

What is a person called who studies insects?

SCIENCE
Quiz Whiz

Grasshoppers and dragonflies go through how many stages of metamorphosis?

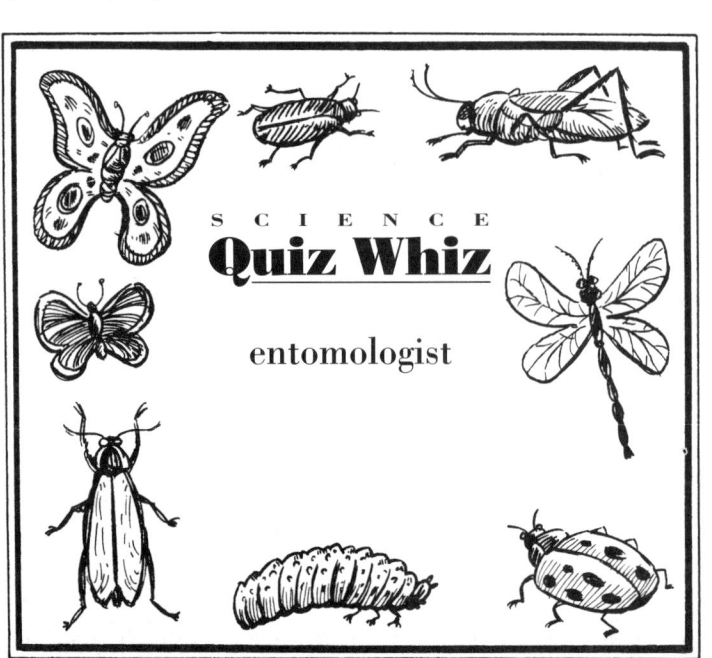

S C I E N C E
Quiz Whiz

Where in an insect's body is food temporarily stored and digested before moving on to the gizzard?

S C I E N C E
Quiz Whiz

What crop does the boll weevil destroy?

S C I E N C E
Quiz Whiz

What is the outside skeleton of an insect called?

S C I E N C E
Quiz Whiz

Which is *not* a part of an insect's head: thorax, mouthparts, eyes, or antennae?

S C I E N C E
Quiz Whiz

What are the grinding jaws of an insect called?

S C I E N C E
Quiz Whiz

What insect produces light in its thorax?

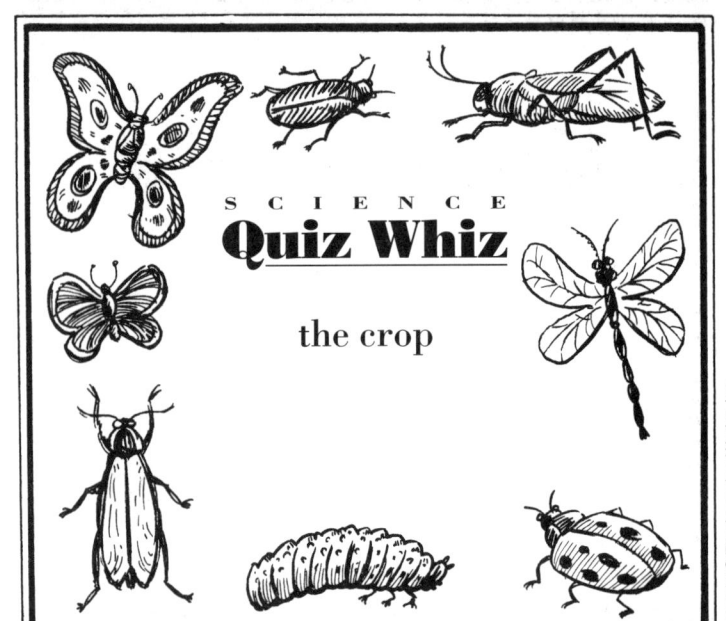

SCIENCE
Quiz Whiz

What is the name of the middle section of an insect's body?

SCIENCE
Quiz Whiz

What part of most insects' bodies is responsible for the sense of smell?

SCIENCE
Quiz Whiz

Which is *not* a stage of metamorphosis: egg, web, larvae, or pupa?

SCIENCE
Quiz Whiz

Which of the following is *not* considered a fly because it has two pairs of wings: fruit fly, gnat, mosquito, or dragonfly?

SCIENCE
Quiz Whiz

What is the main purpose of the chemical light in the abdomen of a firefly: eating, attracting a mate, protection, or flight?

SCIENCE
Quiz Whiz

Which of the following is *not* a type of butterfly: luna, milkweed, monarch, or swallowtail?

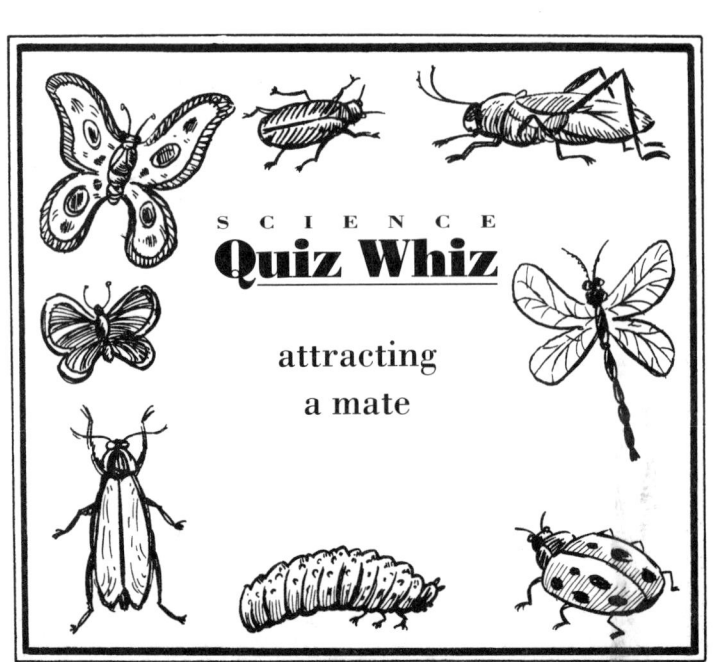

SCIENCE Quiz Whiz

What are animals called that are cold-blooded, have no scales, and live both on land and in the water?

SCIENCE Quiz Whiz

What organs help young amphibians breathe?

SCIENCE Quiz Whiz

What organs do adult amphibians use to breathe?

SCIENCE Quiz Whiz

What is the outer layer of skin called that protects the deeper tissues of amphibians?

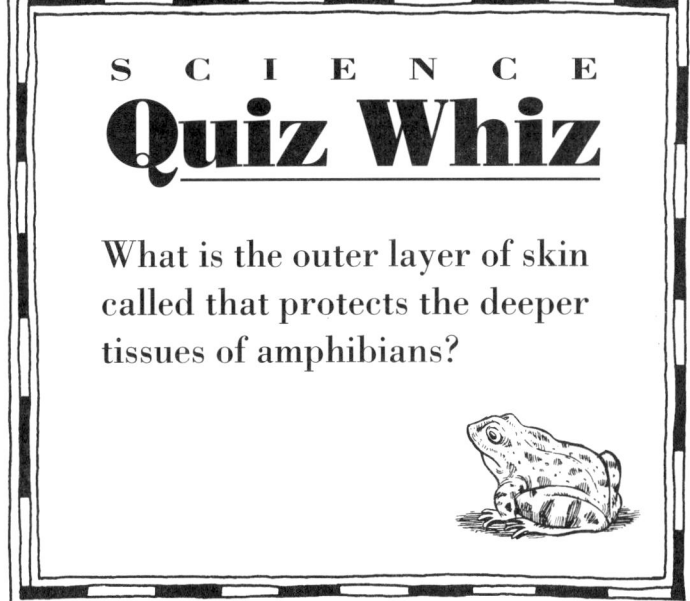

SCIENCE Quiz Whiz

What is the process called by which the larvae of amphibians turn into adults?

SCIENCE Quiz Whiz

Which of the following is an example of an amphibian: eel, salamander, tortoise, or katydid?

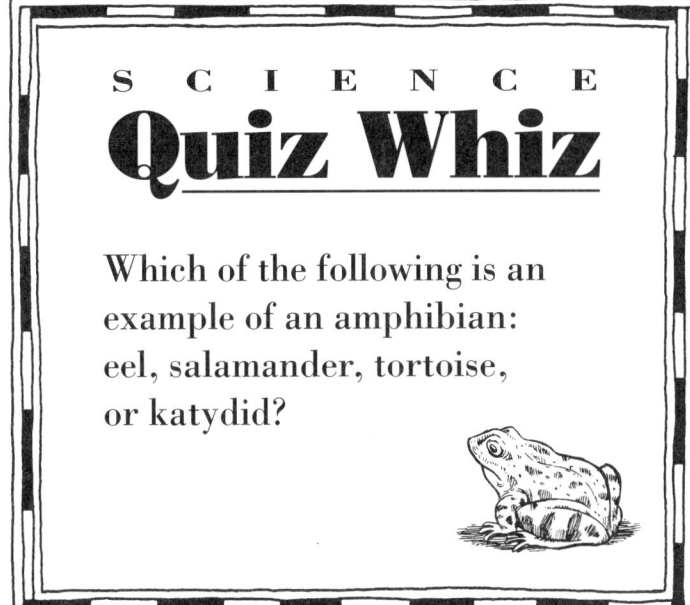

SCIENCE Quiz Whiz

Which of the following does *not* prey on amphibians: birds, snakes, mammals, or insects?

SCIENCE Quiz Whiz

What is the name of the class to which amphibians belong?

SCIENCE Quiz Whiz

Adult frogs and toads do *not* have which of the following: skin, stomach, tail, or liver?

SCIENCE Quiz Whiz

Where do adult amphibians breed: in dirt, in water, in trees, or in nests?

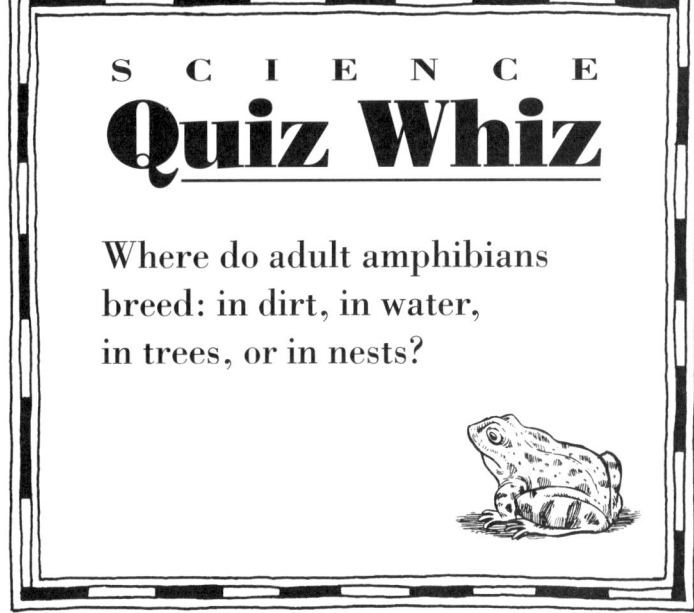

SCIENCE Quiz Whiz

Which is *not* an example of an amphibian: newt, frog, clam, or toad?

SCIENCE Quiz Whiz

What is the fishlike stage called in the life cycle of a frog?

vertebrates

insects

water

tail

tadpole

clam

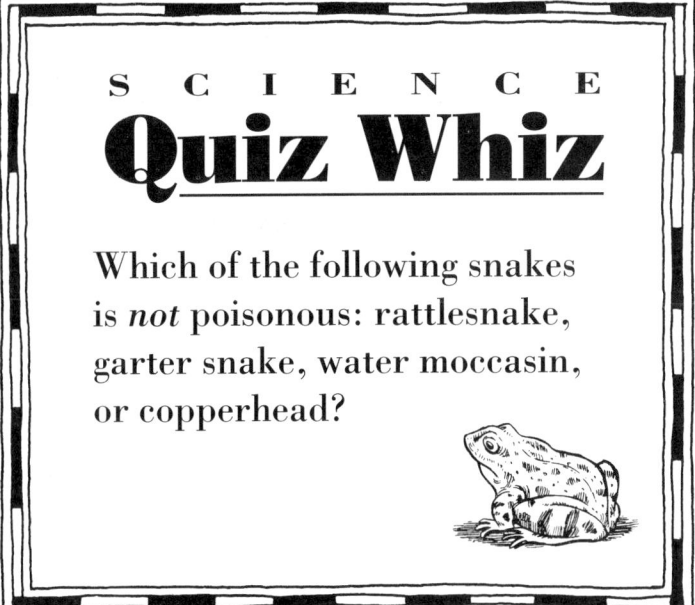

SCIENCE
Quiz Whiz

Which of the following snakes is *not* poisonous: rattlesnake, garter snake, water moccasin, or copperhead?

SCIENCE
Quiz Whiz

What is a cold-blooded vertebrate called that has lungs and dry skin?

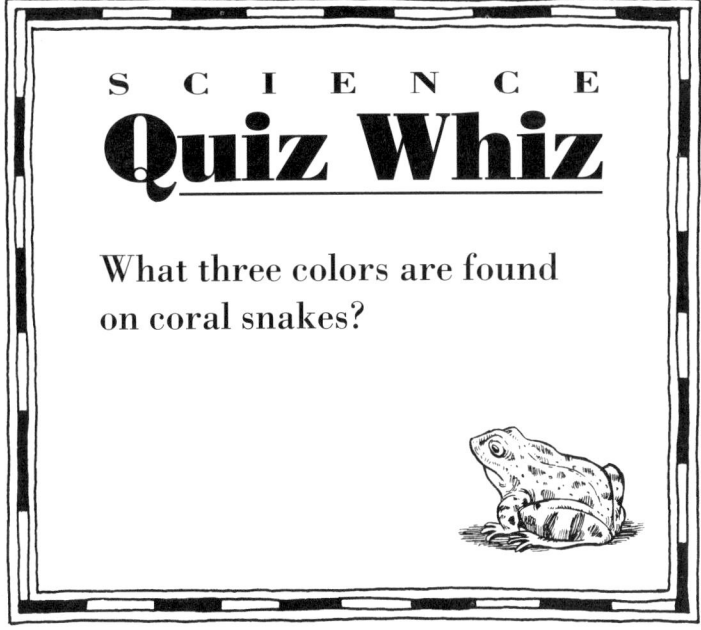

SCIENCE
Quiz Whiz

What three colors are found on coral snakes?

SCIENCE
Quiz Whiz

Is a "terrapin" a type of lizard, snake, turtle, or crocodile?

SCIENCE
Quiz Whiz

Which of the following is *not* a reptile: alligator, turtle, snake, lizard, or shrew?

SCIENCE
Quiz Whiz

What is the skin of a lizard covered with?

reptile

garter snake

turtle

red, yellow, and black

scales

shrew

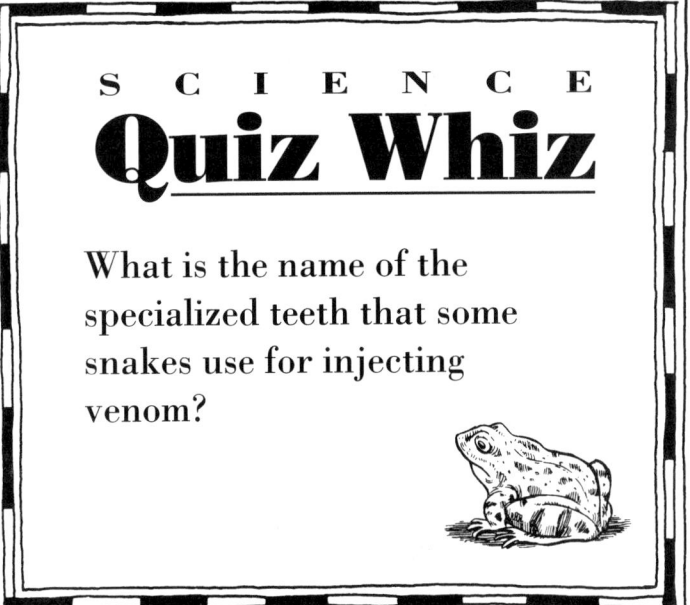

SCIENCE
Quiz Whiz

What is the name of the specialized teeth that some snakes use for injecting venom?

SCIENCE
Quiz Whiz

Is "sidewinder" the name of an alligator, a snake, a toad, or a salamander?

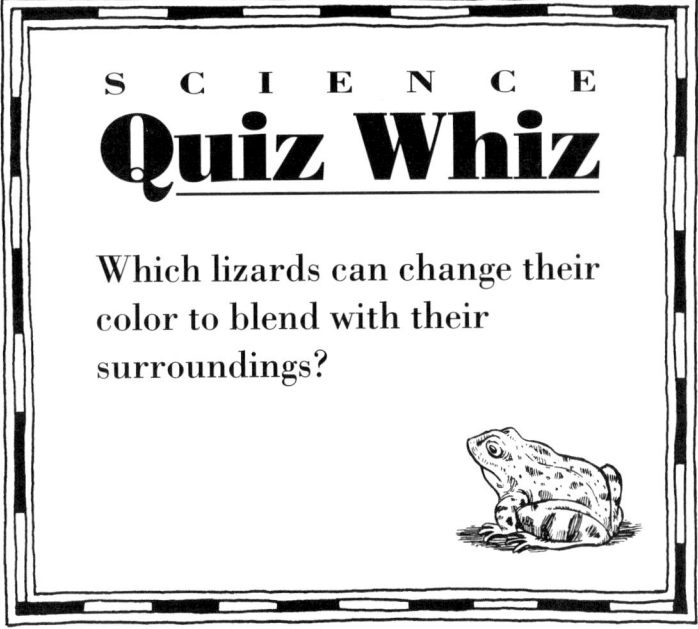

SCIENCE
Quiz Whiz

Which lizards can change their color to blend with their surroundings?

SCIENCE
Quiz Whiz

Which is the longest snake in the world?

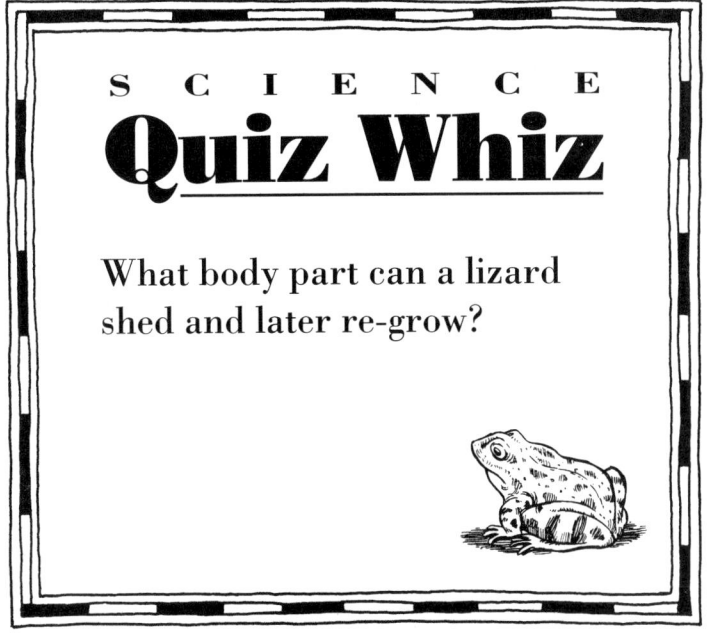

SCIENCE
Quiz Whiz

What body part can a lizard shed and later re-grow?

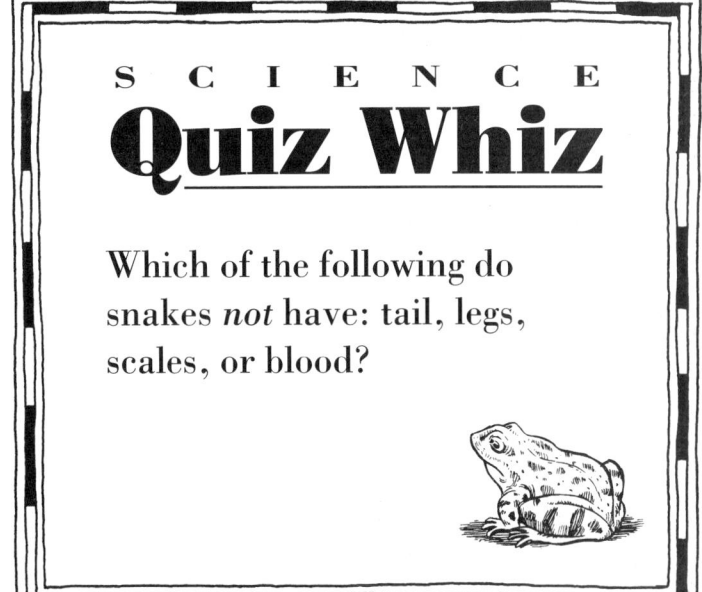

SCIENCE
Quiz Whiz

Which of the following do snakes *not* have: tail, legs, scales, or blood?

SCIENCE Quiz Whiz

What are the only reptiles with shells?

SCIENCE Quiz Whiz

Indigo, hognose, and corn are names of what type of reptile?

SCIENCE Quiz Whiz

The Gila monster is what type of reptile: lizard, snake, alligator, or turtle?

SCIENCE Quiz Whiz

Which snake does *not* suffocate its prey by constriction: king snake, python, garter snake, or rat snake?

SCIENCE Quiz Whiz

Reptiles breathe by means of which organs?

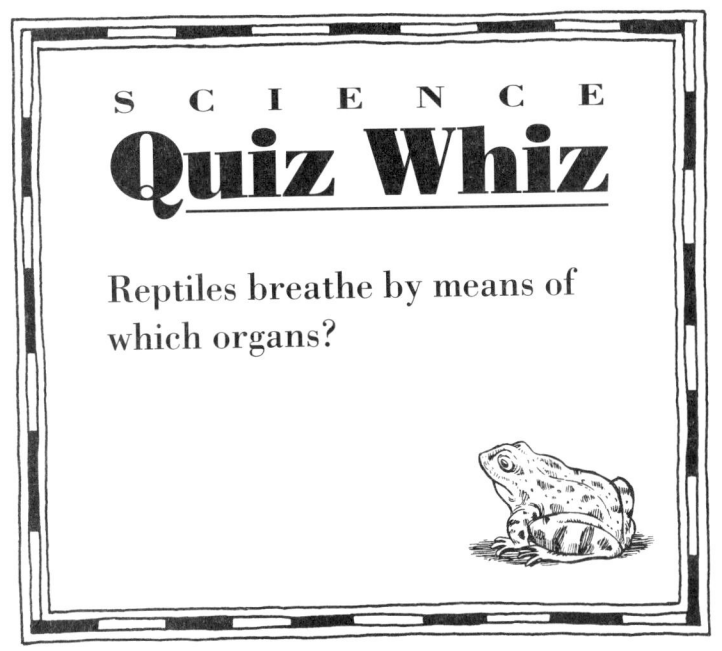

SCIENCE Quiz Whiz

Which snake forms a hood with its skin when it is disturbed?

SCIENCE
Quiz Whiz
snake

SCIENCE
Quiz Whiz
turtles and tortoises

SCIENCE
Quiz Whiz
garter snake

SCIENCE
Quiz Whiz
lizard

SCIENCE
Quiz Whiz
cobra

SCIENCE
Quiz Whiz
lungs

SCIENCE Quiz Whiz

Which of the following is another name for the shell of a turtle: carapace, sac, barbell, or embryo?

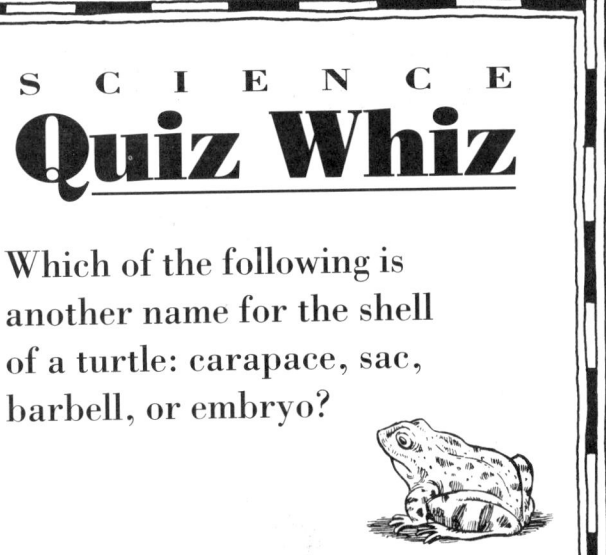

SCIENCE Quiz Whiz

What does a reptile do when it molts?

SCIENCE Quiz Whiz

Which reptiles can move their upper and lower jaws apart when they swallow?

SCIENCE Quiz Whiz

Which is *not* a type of turtle: terrapin, mollusk, loggerhead, or leatherback?

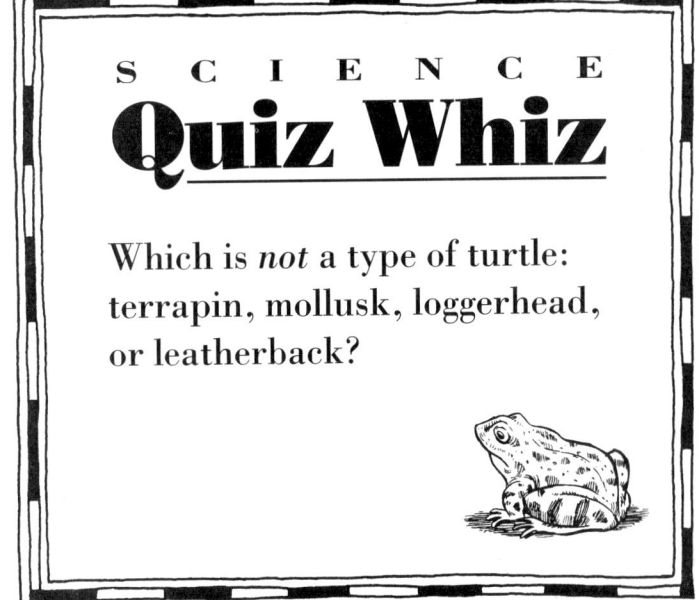

SCIENCE Quiz Whiz

Which of the following do turtles *not* have: teeth, tail, shell, or lungs?

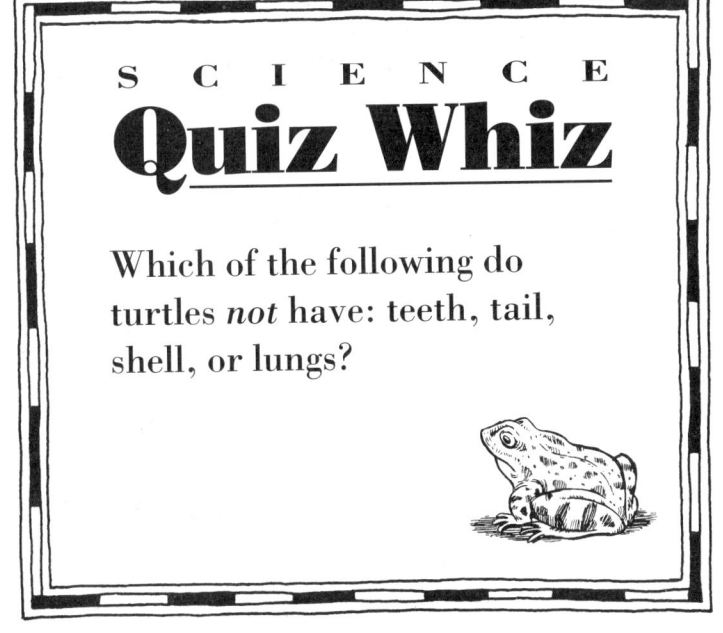

SCIENCE Quiz Whiz

Which is *not* a kind of lizard: chameleon, gecko, iguana, or viper?

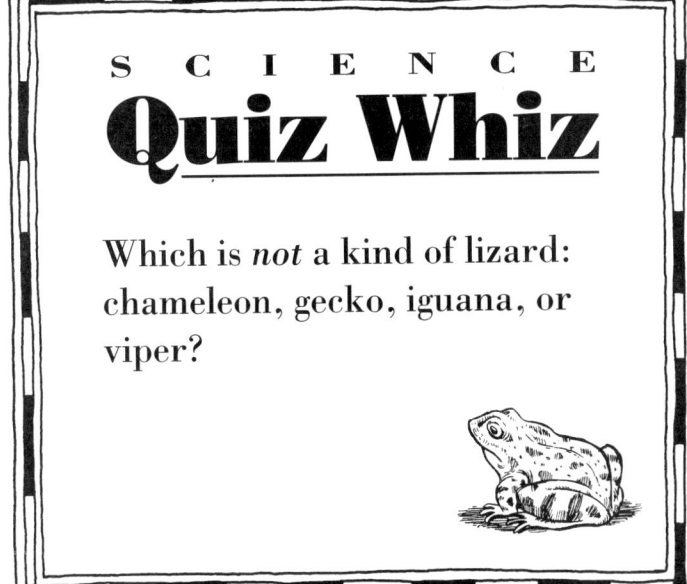

SCIENCE
Quiz Whiz

sheds its skin

SCIENCE
Quiz Whiz

carapace

SCIENCE
Quiz Whiz

mollusk

SCIENCE
Quiz Whiz

snakes

SCIENCE
Quiz Whiz

viper

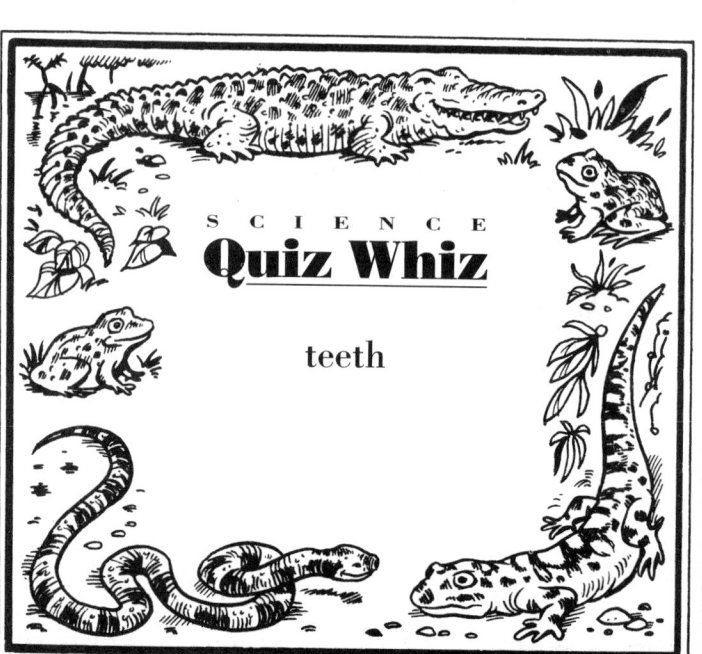

SCIENCE
Quiz Whiz

teeth

SCIENCE
Quiz Whiz

Which bird can fly backwards and hover in mid-air?

SCIENCE
Quiz Whiz

What is the largest bird in the world?

SCIENCE
Quiz Whiz

What is a scientist called who studies birds?

SCIENCE
Quiz Whiz

What term is given to birds, such as owls, that are active at night?

SCIENCE
Quiz Whiz

Which bird would you most likely find in a desert: flamingo, egret, cactus wren, or goldfinch?

SCIENCE
Quiz Whiz

What is the national bird of the United States?

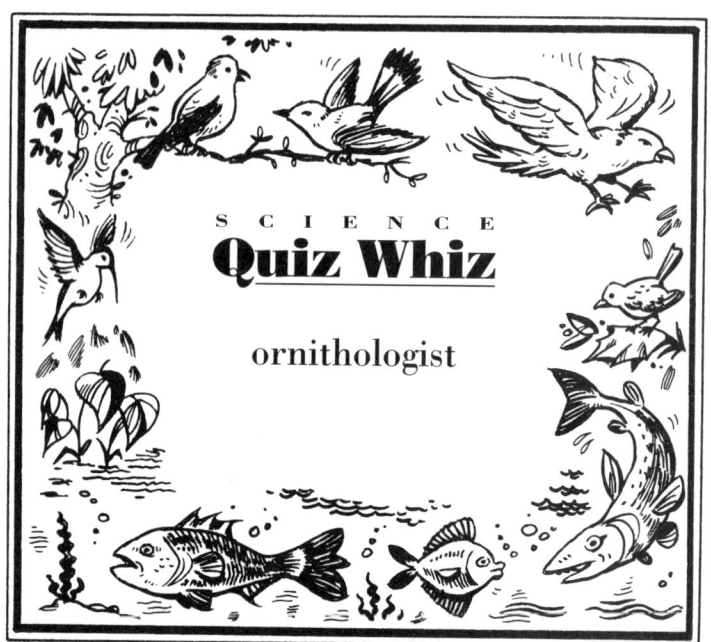

SCIENCE
Quiz Whiz

Which is *not* a kind of bird: finch, quetzal, vulture, or lamprey?

SCIENCE
Quiz Whiz

What term describes an animal whose body temperature changes with the temperature of the air or water around it?

SCIENCE
Quiz Whiz

Which bird would you most likely find at a lake: heron, robin, cactus wren, or hummingbird?

SCIENCE
Quiz Whiz

Which is *not* a bird of prey: pheasant, eagle, hawk, or owl?

SCIENCE
Quiz Whiz

What distinctive feature do most quails have: a curved tail, a head plume, a star pattern on the breast, or red tail feathers?

SCIENCE
Quiz Whiz

What male bird is noted for its bright red color: oriole, cardinal, pelican, or roadrunner?

SCIENCE
Quiz Whiz
cold-blooded or exothermic

SCIENCE
Quiz Whiz
lamprey

SCIENCE
Quiz Whiz
pheasant

SCIENCE
Quiz Whiz
heron

SCIENCE
Quiz Whiz
cardinal

SCIENCE
Quiz Whiz
plume

SCIENCE
Quiz Whiz

Which of the following is *not* a fish: barracuda, marlin, dolphin, or pike?

SCIENCE
Quiz Whiz

What is a baby ostrich called?

SCIENCE
Quiz Whiz

Which bird often lays its eggs in other birds' nests: crane, parakeet, hornbill, or cuckoo?

SCIENCE
Quiz Whiz

Which of the following birds has webbed feet: parrot, pelican, toucan, or egret?

SCIENCE
Quiz Whiz

What do ostriches, penguins, and emus have in common?

SCIENCE
Quiz Whiz

What is the smallest bird in the world, measuring about two inches from beak to tail?

SCIENCE
Quiz Whiz
chick

SCIENCE
Quiz Whiz
dolphin

SCIENCE
Quiz Whiz
pelican

SCIENCE
Quiz Whiz
cuckoo

SCIENCE
Quiz Whiz
bee hummingbird

SCIENCE
Quiz Whiz
they can't fly

SCIENCE
Quiz Whiz

A macaw is what type of bird: parrot, bluebird, cardinal, or duck?

SCIENCE
Quiz Whiz

What color is a robin's egg?

SCIENCE
Quiz Whiz

Orioles are noted for which two distinctive colors: green and red, orange and black, purple and yellow, or red and blue?

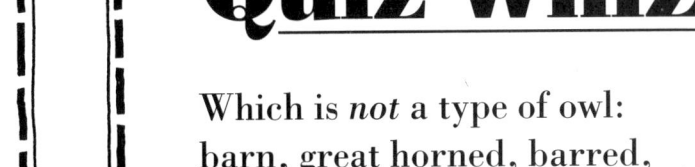

SCIENCE
Quiz Whiz

Which is *not* a type of owl: barn, great horned, barred, or bearded?

SCIENCE
Quiz Whiz

Which of the following is *not* a type of crustacean: lobster, salmon, crayfish, or shrimp?

SCIENCE
Quiz Whiz

What structure helps fish move through the water and maintain balance?

SCIENCE
Quiz Whiz

Which fish does *not* live in fresh water: trout, perch, mackerel, sunfish, salmon, or bass?

SCIENCE
Quiz Whiz

Instead of bone, what is the skeleton of a shark made of?

SCIENCE
Quiz Whiz

Which of the following fishes can live in both saltwater and fresh water: goldfish, salmon, tuna, or dolphins?

SCIENCE
Quiz Whiz

What kind of a fish is a hammerhead?

SCIENCE
Quiz Whiz

What is a baby fish called?

SCIENCE
Quiz Whiz

Which of the following is *not* really a fish: jellyfish, sailfish, guppy, or filefish?

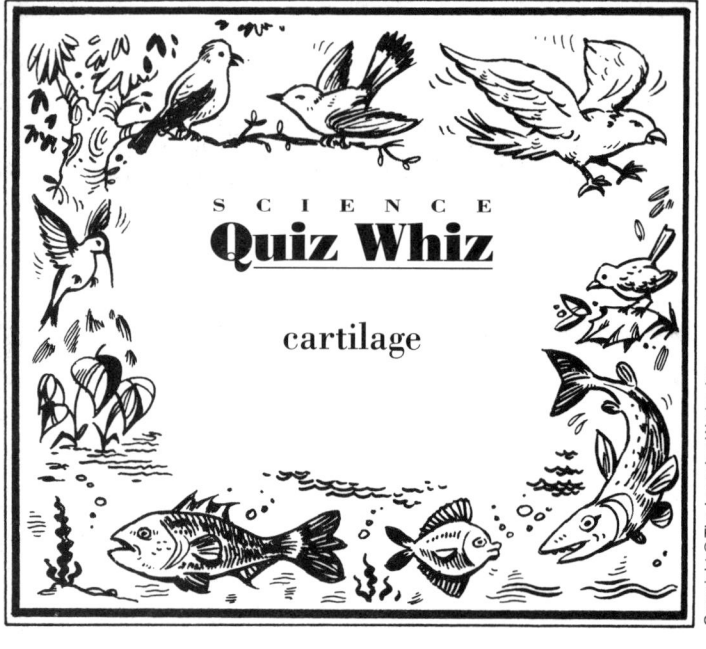

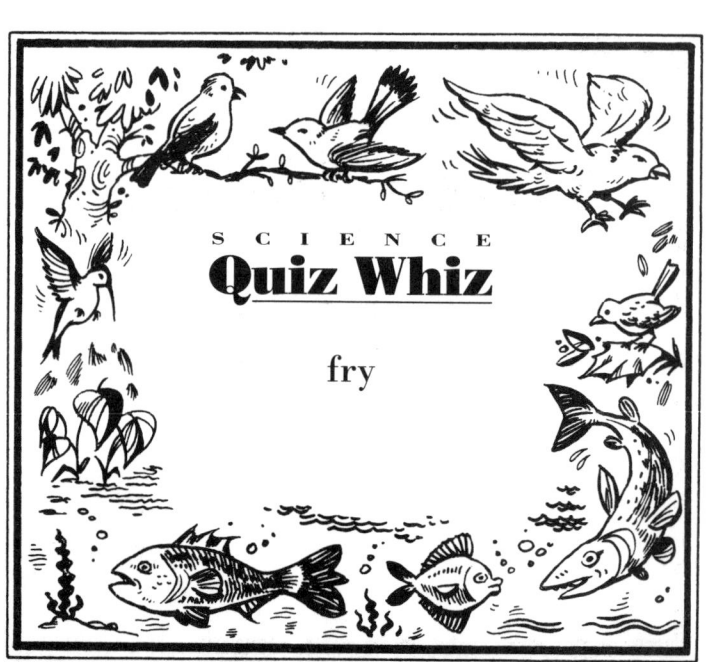

SCIENCE Quiz Whiz

Which fish carries its young in a pouch like a kangaroo?

SCIENCE Quiz Whiz

What type of flat fish is related to the shark and has eyes on the top of its head?

SCIENCE Quiz Whiz

The heart of a fish has how many chambers?

SCIENCE Quiz Whiz

What is a scientist called who studies fish?

SCIENCE Quiz Whiz

Which is the shark's most acute sense: sight, taste, or smell?

SCIENCE Quiz Whiz

A goldfish is a miniature form of what other fish?

SCIENCE
Quiz Whiz
ray

SCIENCE
Quiz Whiz
seahorse

SCIENCE
Quiz Whiz
ichthyologist

SCIENCE
Quiz Whiz
two chambers

SCIENCE
Quiz Whiz
carp

SCIENCE
Quiz Whiz
smell

SCIENCE Quiz Whiz

What name is given to the group of animals that have spinal cords encased in backbones?

SCIENCE Quiz Whiz

What is the smallest breed of dog in the world?

SCIENCE Quiz Whiz

What is the name of the thick layer of fat found on whales and dolphins that helps keep them warm in cold water?

SCIENCE Quiz Whiz

What is the term for an animal that hunts other animals for food?

SCIENCE Quiz Whiz

What are animals with pouches called?

SCIENCE Quiz Whiz

What is a baby deer called?

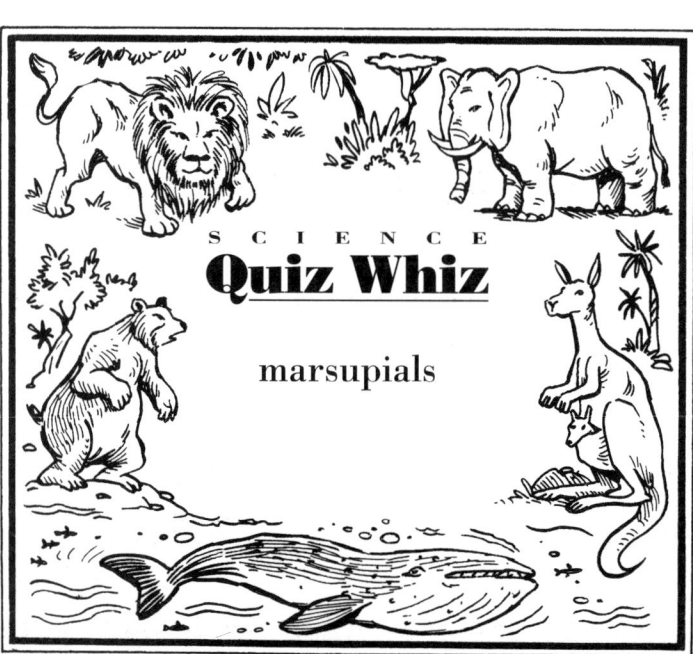

SCIENCE Quiz Whiz

In the classification of animals, which is the smallest group: species, order, class, or genus?

SCIENCE Quiz Whiz

Which of the following is *not* a type of bear: red, grizzly, polar, or spectacled?

SCIENCE Quiz Whiz

What name is given to a group of elephants?

SCIENCE Quiz Whiz

What name is given to warm-blooded vertebrates whose bodies are usually covered with hair or fur?

SCIENCE Quiz Whiz

What term is applied to animals that are active during the day?

SCIENCE Quiz Whiz

What is a group of lions called?

SCIENCE Quiz Whiz

red

SCIENCE Quiz Whiz

species

SCIENCE Quiz Whiz

mammal

SCIENCE Quiz Whiz

herd

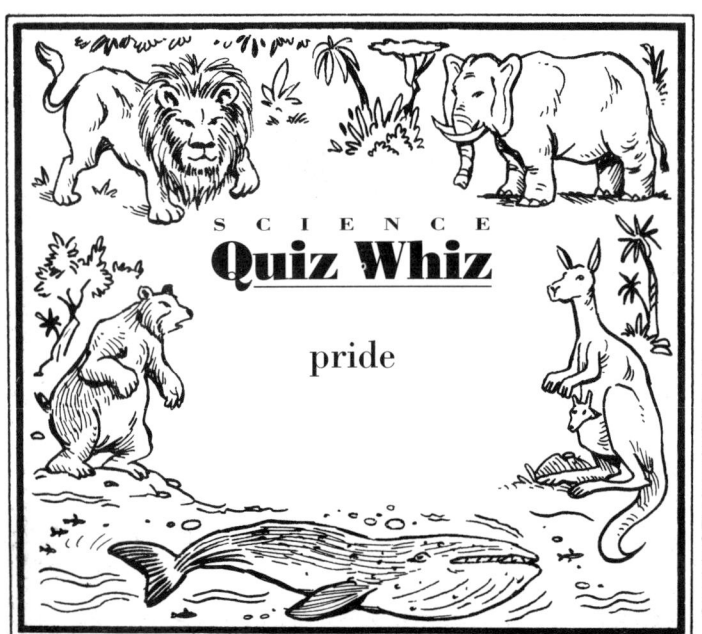

SCIENCE Quiz Whiz

pride

SCIENCE Quiz Whiz

diurnal

SCIENCE
Quiz Whiz

Which is *not* a type of whale: narwhal, orca, beluga, or barracuda?

SCIENCE
Quiz Whiz

What is a group of elk called?

SCIENCE
Quiz Whiz

What is the name of the process by which some mammals periodically shed their hair?

SCIENCE
Quiz Whiz

In the classification of animals, which is the largest group: family, phylum, kingdom, or genus?

SCIENCE
Quiz Whiz

What is a male deer called?

SCIENCE
Quiz Whiz

Which are *not* considered primates: monkeys, apes, human beings, or tigers?

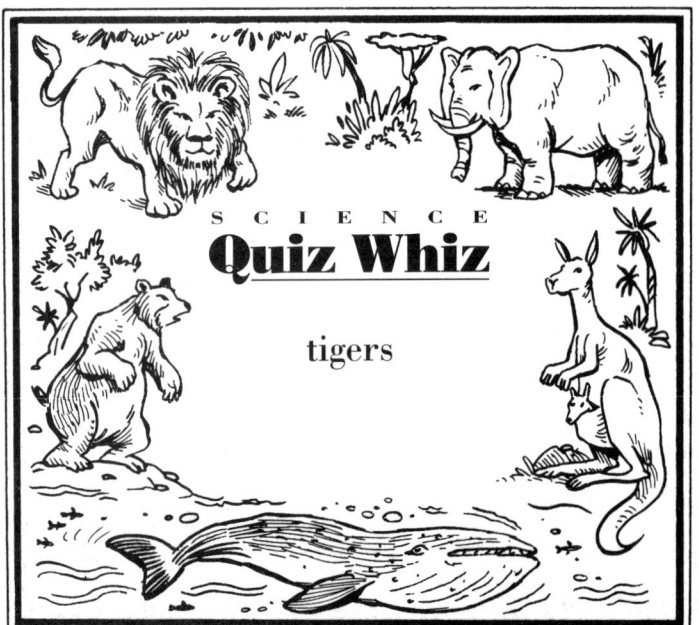

SCIENCE
Quiz Whiz

What name is given to gnawing mammals such as mice, rats, beavers, and porcupines?

SCIENCE
Quiz Whiz

What is a female lion called?

SCIENCE
Quiz Whiz

On what part of a whale are the flukes located?

SCIENCE
Quiz Whiz

What is an animal called whose normal diet includes both plants and animals?

SCIENCE
Quiz Whiz

What is a group of seals called?

SCIENCE
Quiz Whiz

What is a scientist called who studies animals and their classifications?

SCIENCE
Quiz Whiz

Which of the following animals does *not* have horns: goat, elephant, buffalo, or antelope?

SCIENCE
Quiz Whiz

Which of the following animals does *not* have antlers: zebra, caribou, deer, or moose?

SCIENCE
Quiz Whiz

What is a male chicken called?

SCIENCE
Quiz Whiz

What land animal can run the fastest?

SCIENCE
Quiz Whiz

What term is given to animals that stand on four feet, such as elephants and dogs?

SCIENCE
Quiz Whiz

What are the organs called through which animals obtain oxygen from the air?

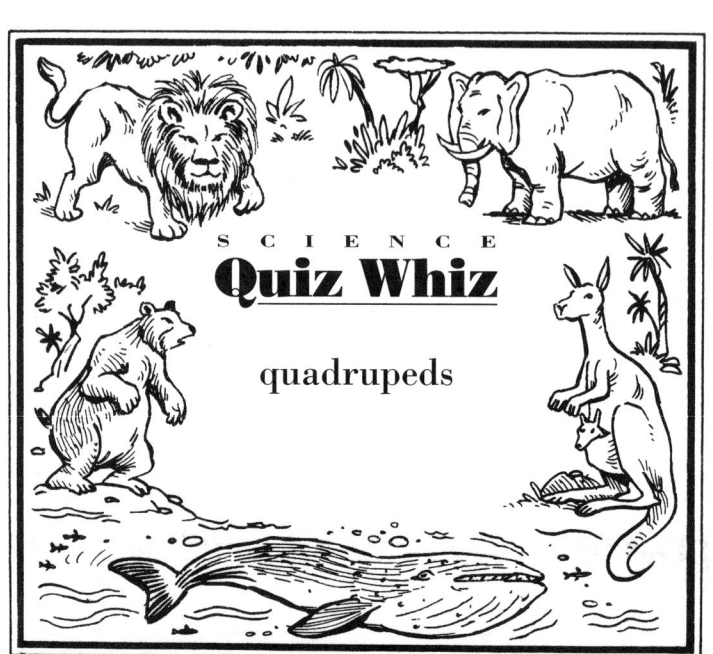

SCIENCE Quiz Whiz

What is the largest animal in the world today?

SCIENCE Quiz Whiz

What is a baby kangaroo called?

SCIENCE Quiz Whiz

What term is given to animals that stand and walk on two legs, such as gorillas?

SCIENCE Quiz Whiz

Which of the following is *not* an herbivore, or plant-eater: cow, lion, sheep, or horse?

SCIENCE Quiz Whiz

In biology, what classification is lower than a class but higher than a family?

SCIENCE Quiz Whiz

Which of the following does *not* hibernate: woodchuck, ground squirrel, elephant, or bear?

SCIENCE Quiz Whiz

What is the name of the smallest units or building blocks of all living things?

SCIENCE Quiz Whiz

How many permanent teeth do most humans have?

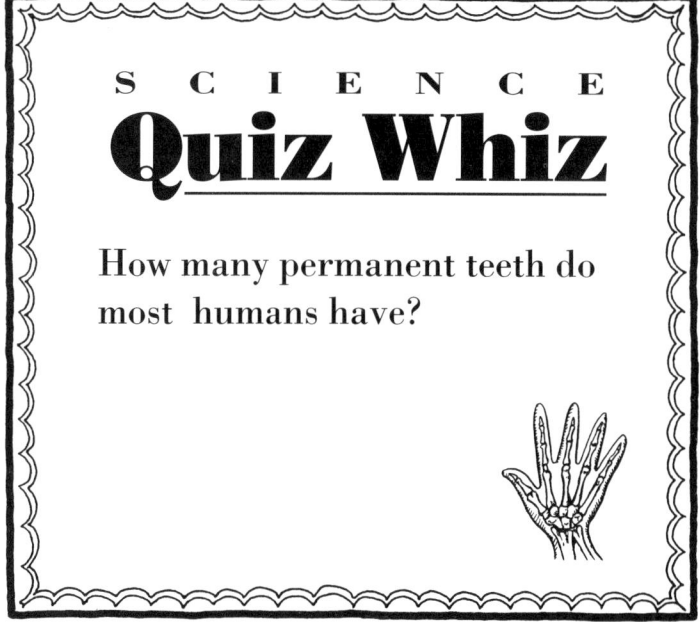

SCIENCE Quiz Whiz

What is another name for the tympanic membrane?

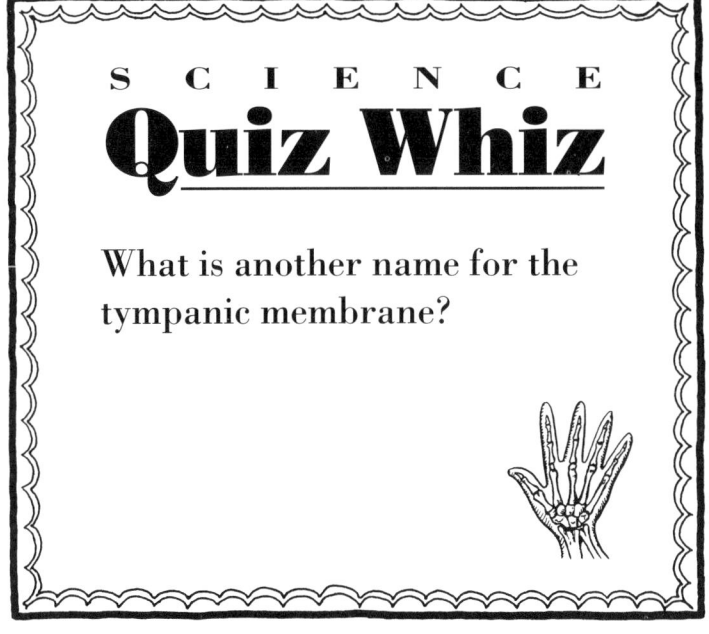

SCIENCE Quiz Whiz

What is the name of the lower part of the large intestine where waste is stored until it is passed from the body?

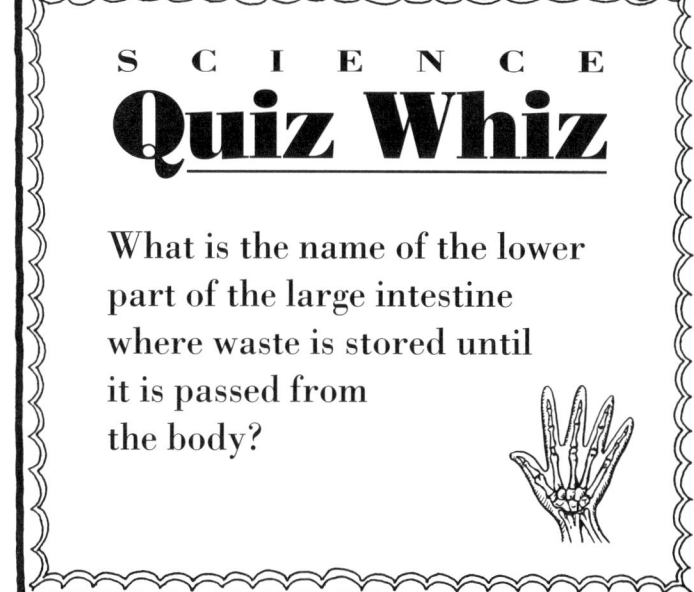

SCIENCE Quiz Whiz

What light-sensitive cells in the eye detect color?

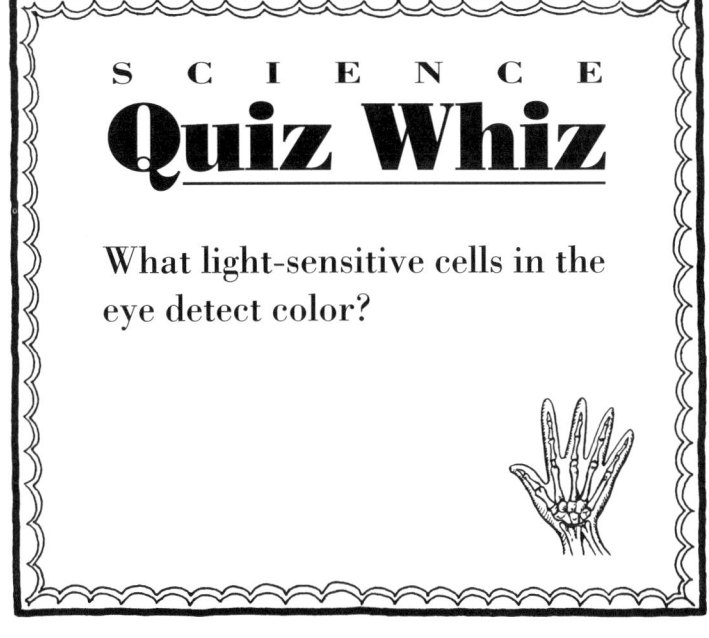

SCIENCE Quiz Whiz

What name is given to the study of the parts of the human body and their functions?

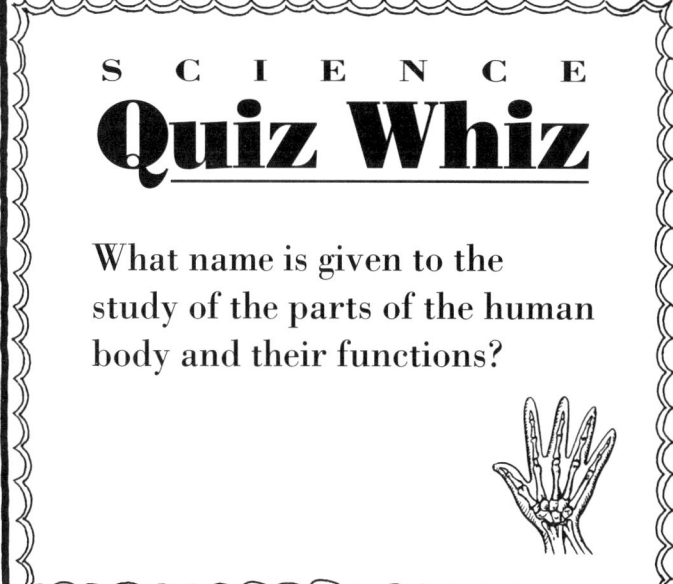

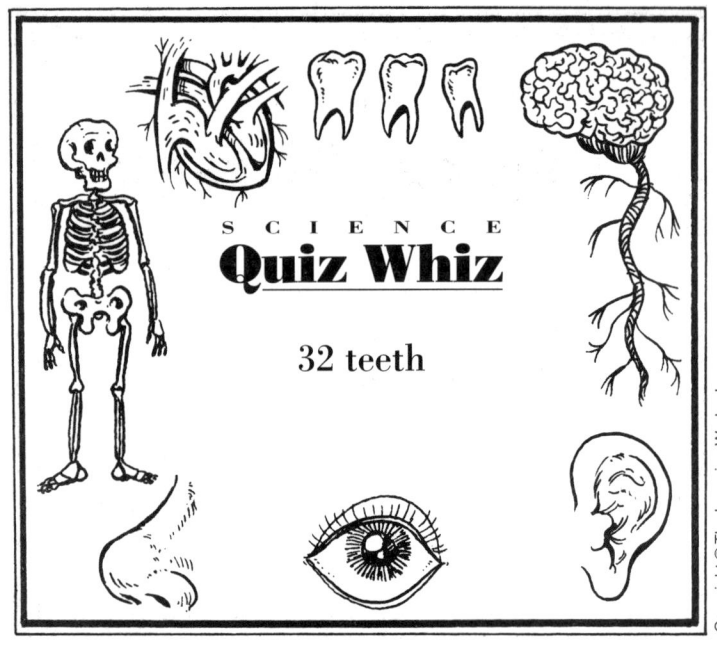

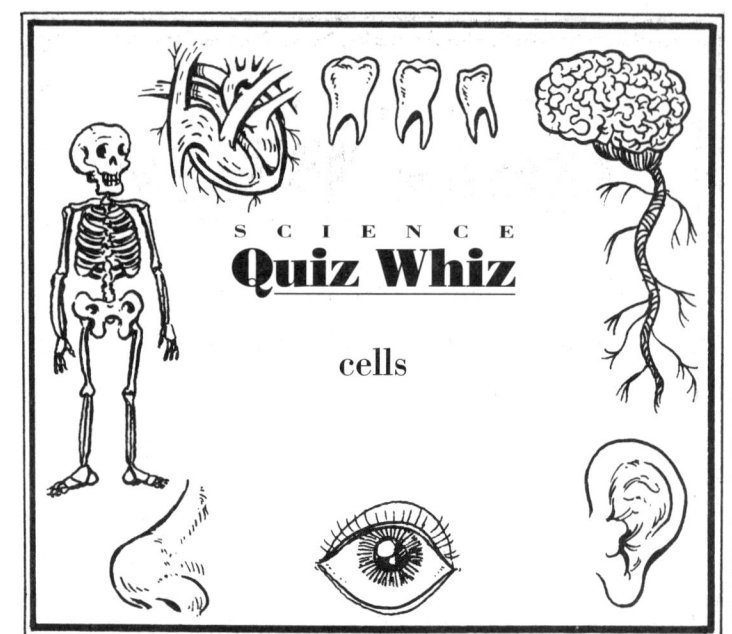

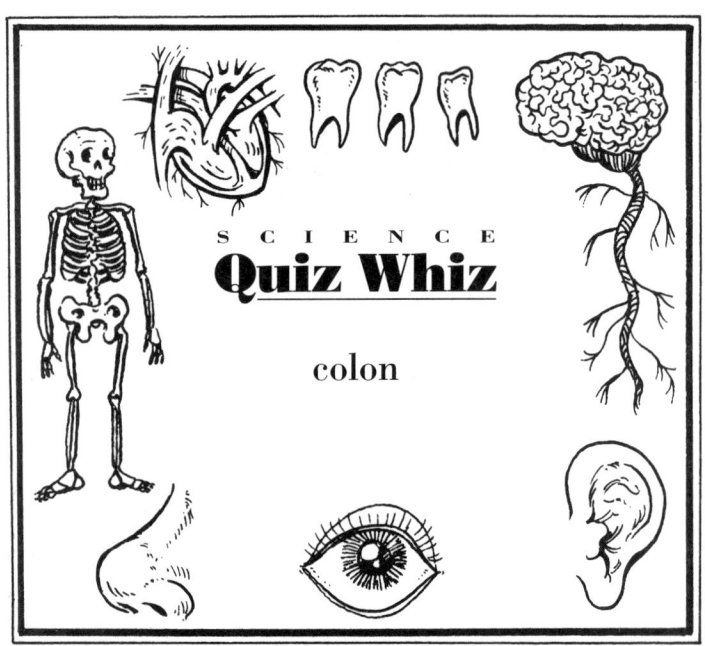

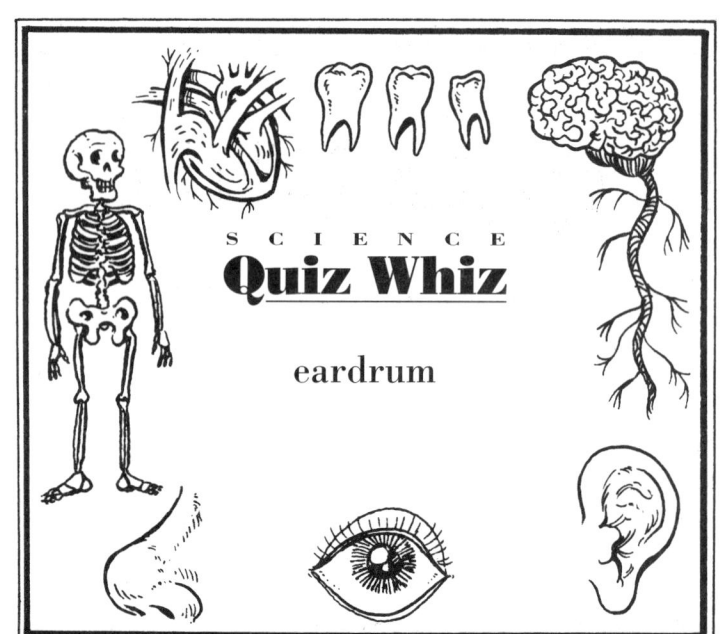

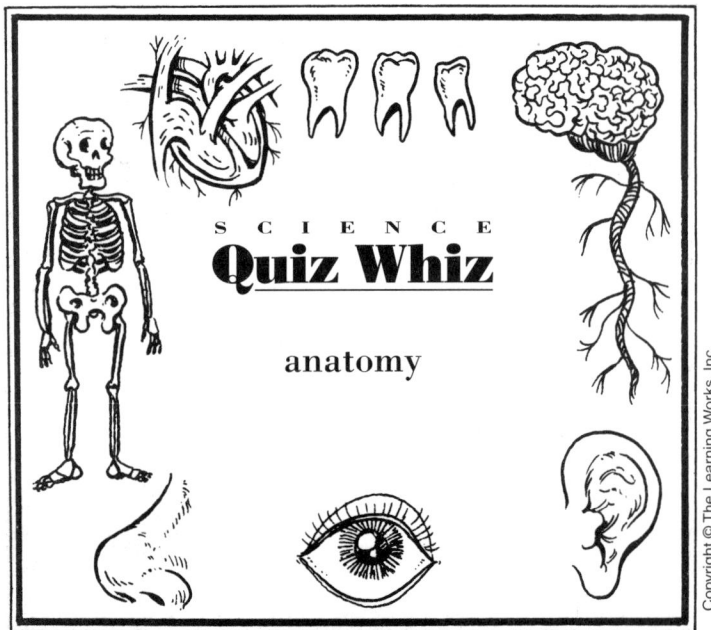

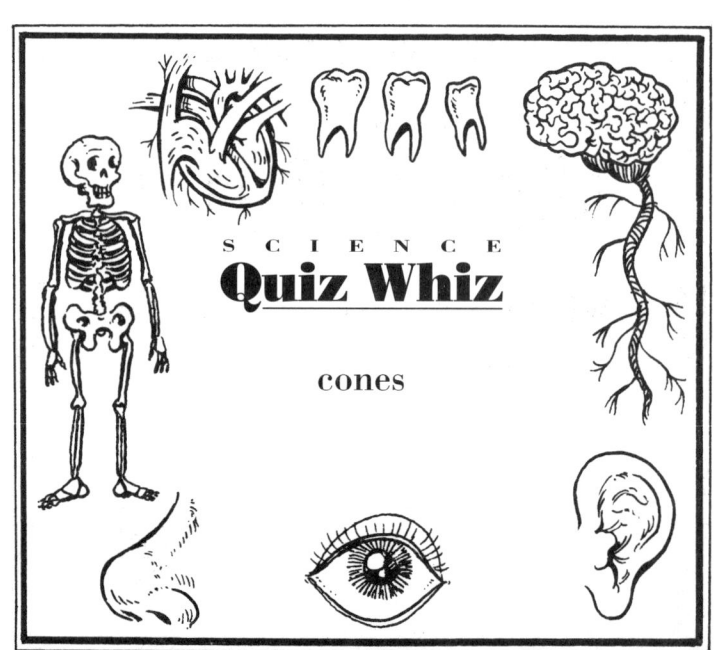

SCIENCE Quiz Whiz

What is the name of the tough tissue that connects muscles to bones?

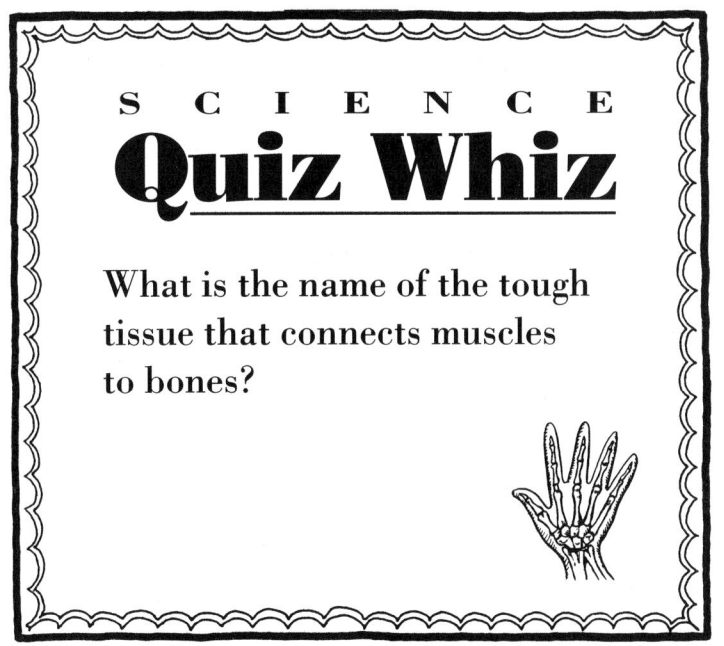

SCIENCE Quiz Whiz

What is the name of the narrow tube that connects the middle ear to the throat?

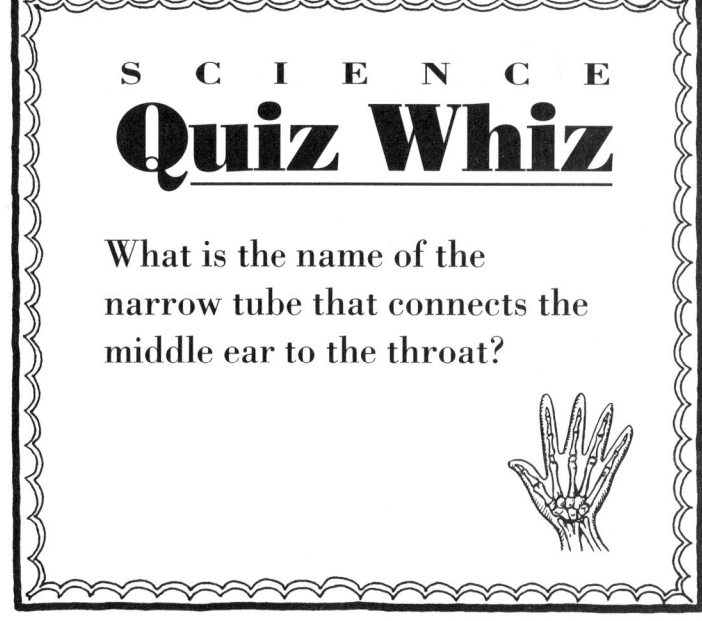

SCIENCE Quiz Whiz

What is the inner layer of skin called?

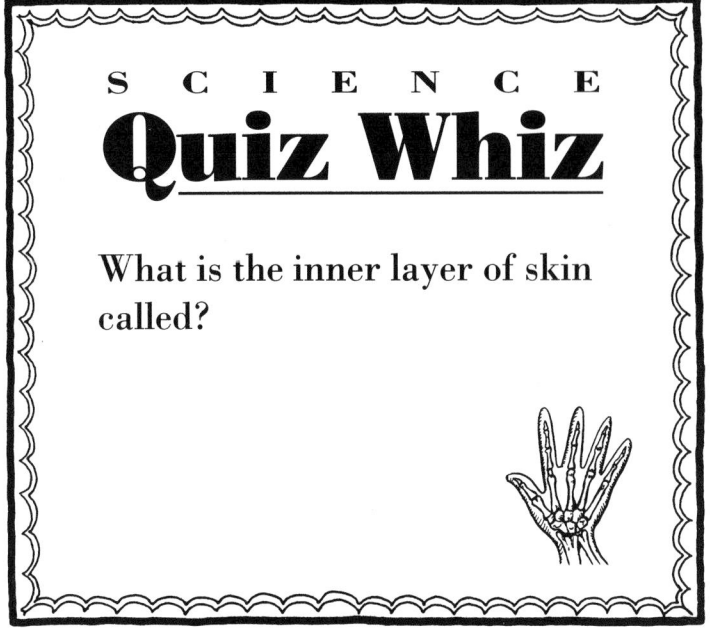

SCIENCE Quiz Whiz

What are the sharp, front teeth called that are used for cutting and biting food?

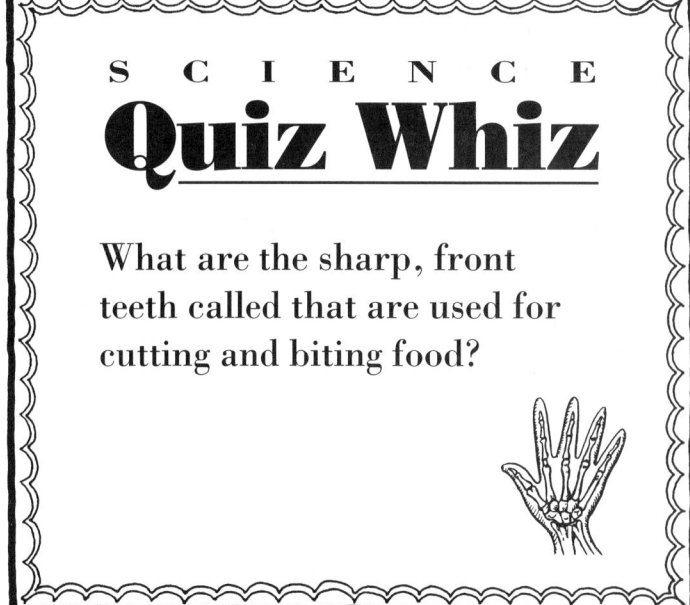

SCIENCE Quiz Whiz

What is the colored part of the eye called?

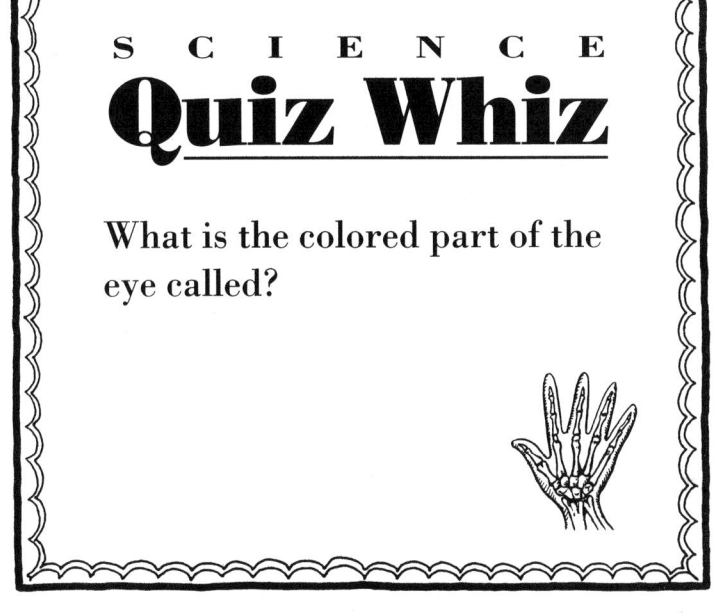

SCIENCE Quiz Whiz

What part of the brain, located at the base of the cerebrum, controls and coordinates voluntary movements?

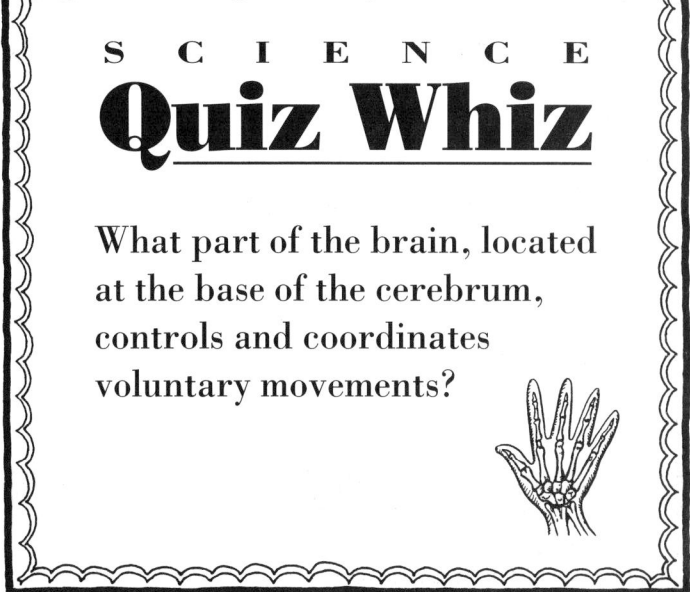

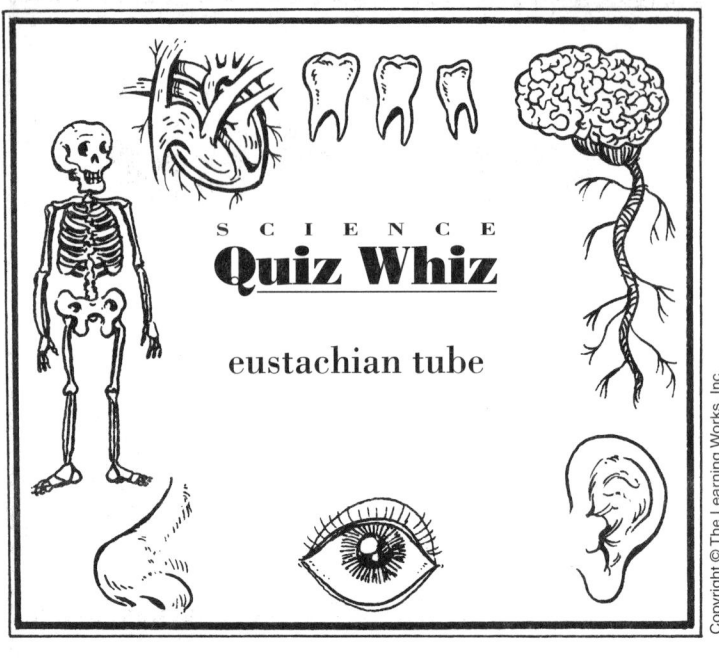
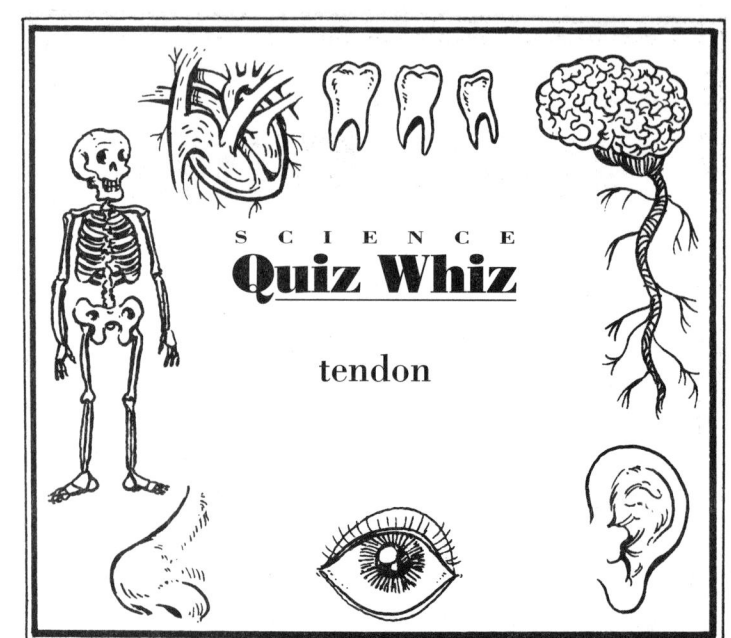
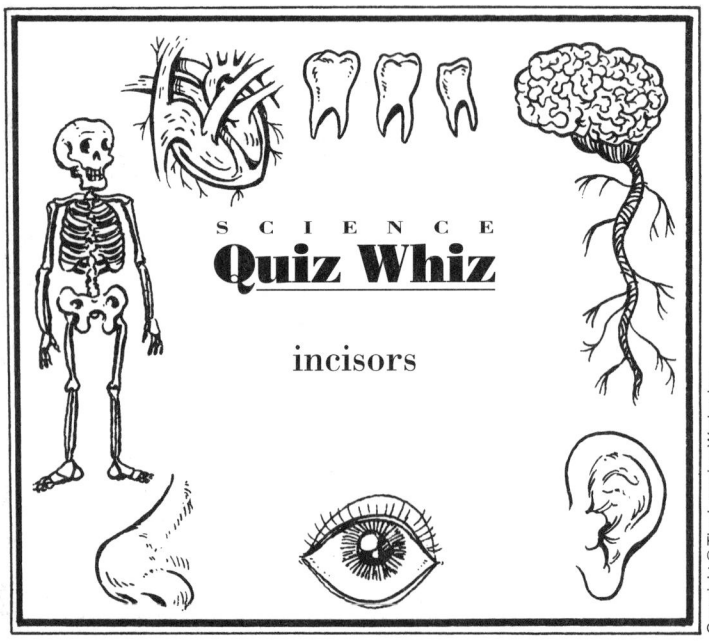
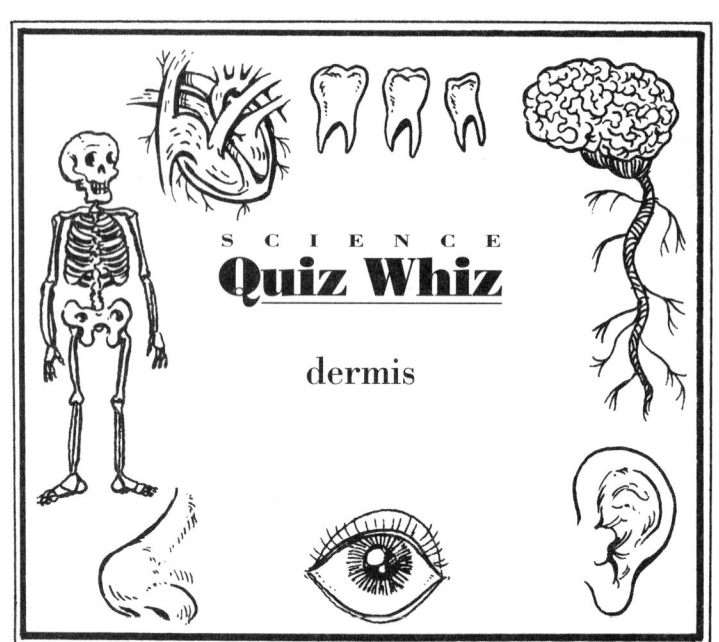
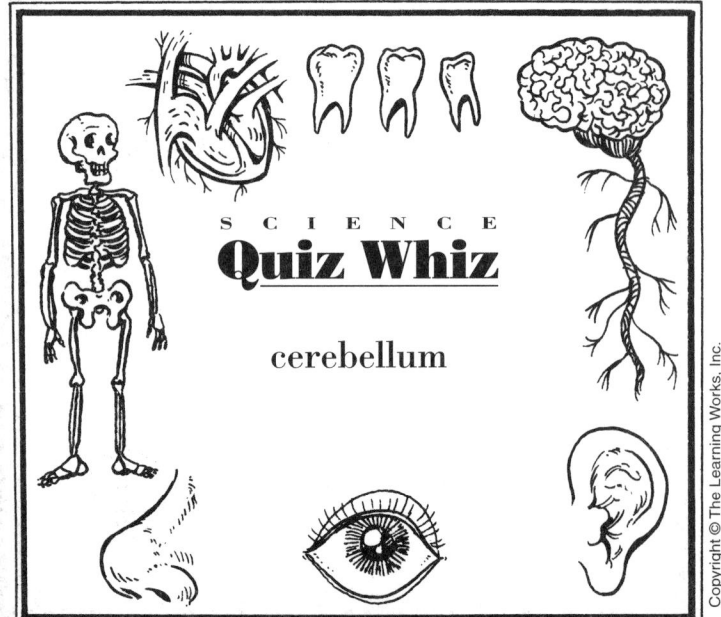
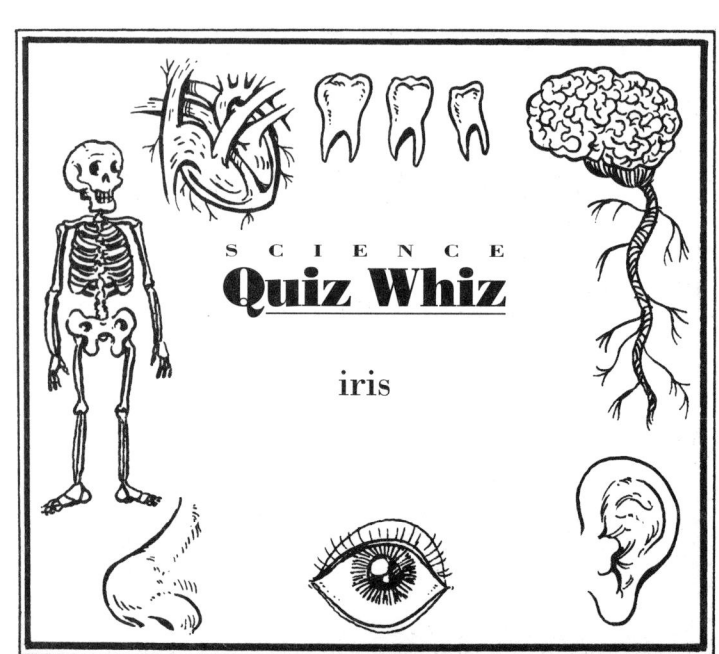

SCIENCE Quiz Whiz

What is the name of the blood vessels that carry oxygen-rich blood away from the heart?

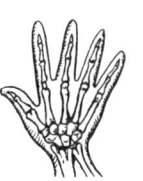

SCIENCE Quiz Whiz

What is another name for the windpipe?

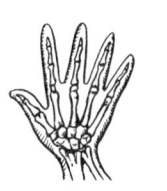

SCIENCE Quiz Whiz

What is the name of the place where one bone connects to another?

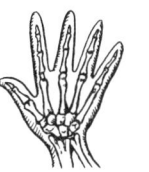

SCIENCE Quiz Whiz

What is the name of the main nerve connecting the eye to the brain?

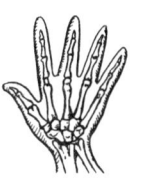

SCIENCE Quiz Whiz

How many bones make up the adult human body?

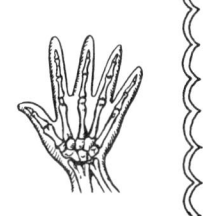

SCIENCE Quiz Whiz

What are the blood vessels called that carry blood to the heart?

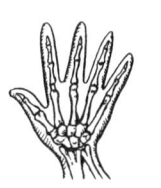

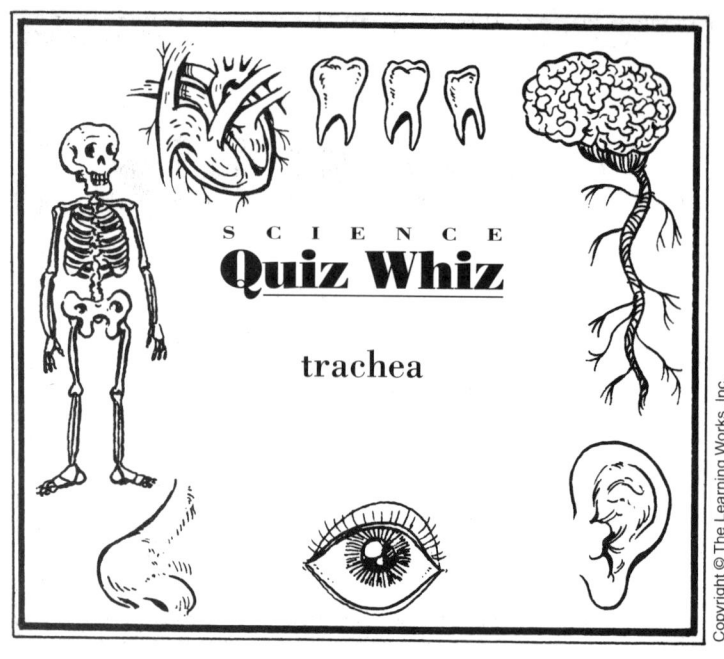
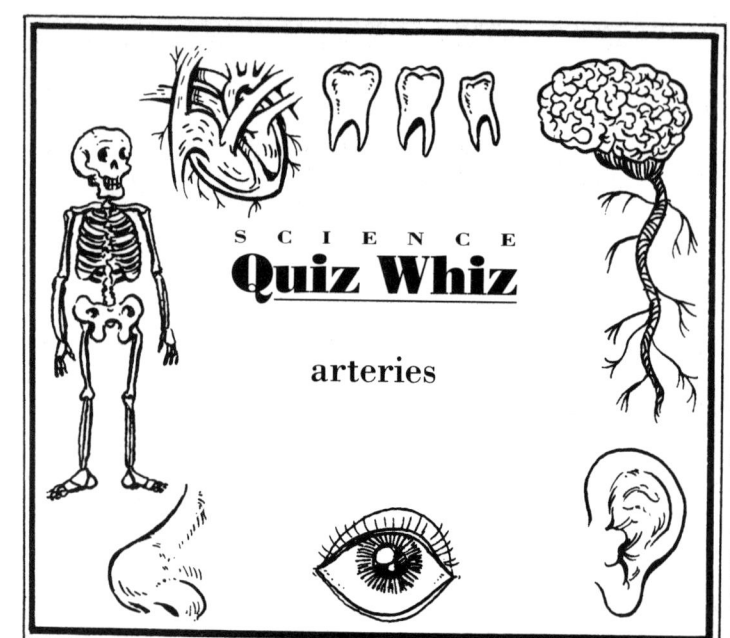
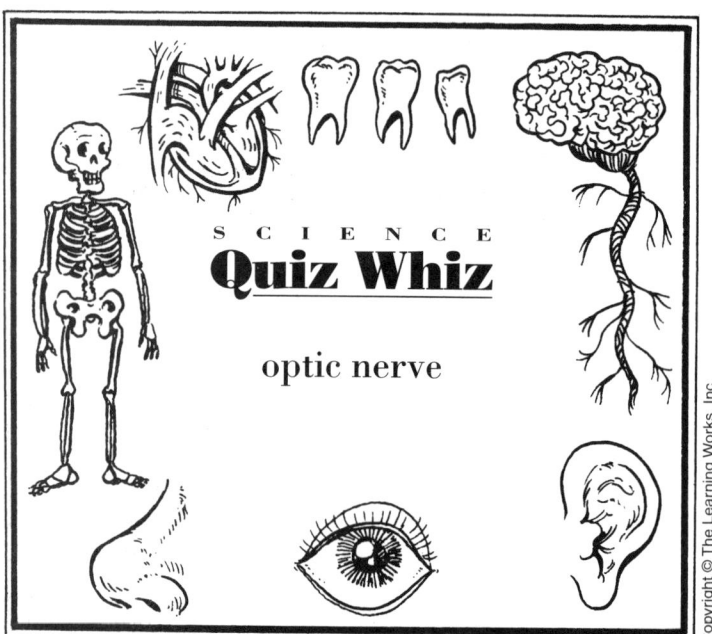
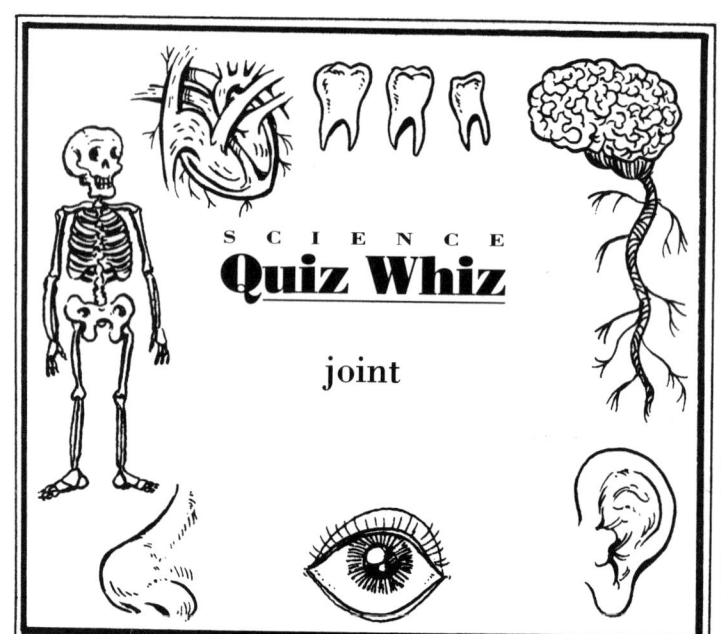
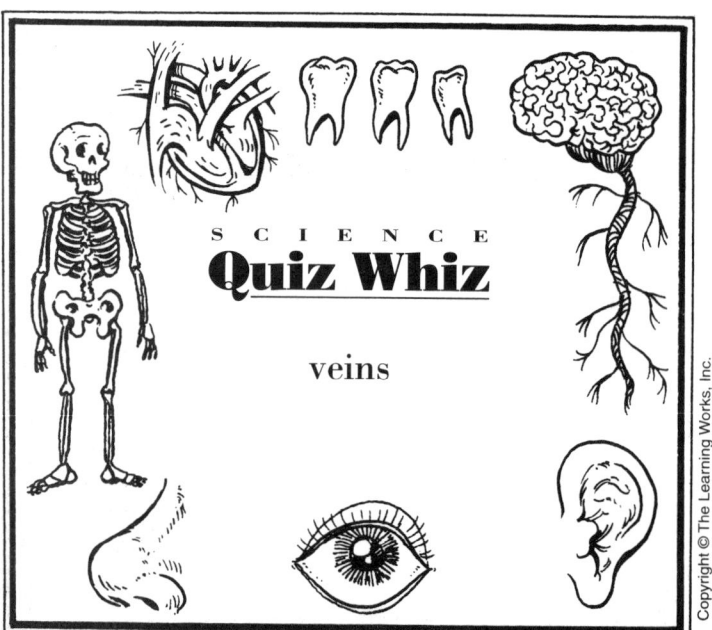
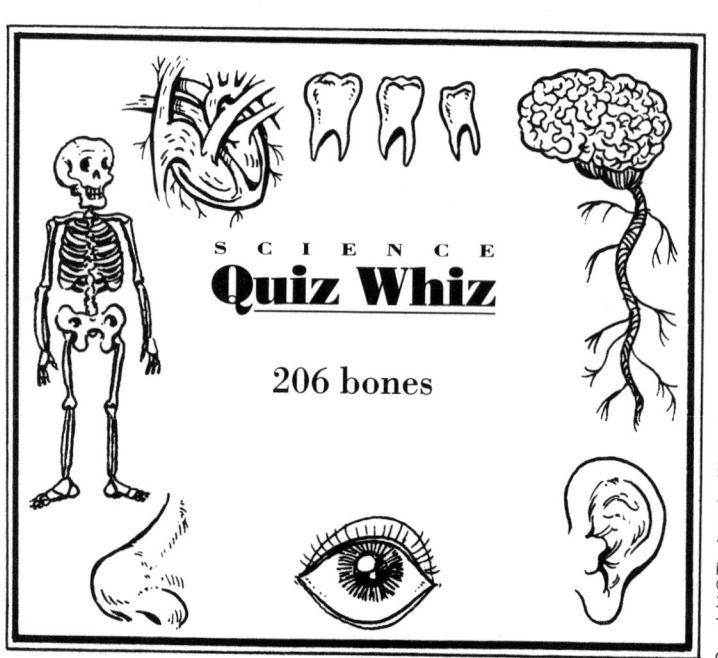

SCIENCE Quiz Whiz

What is the name of the thin outer layer of skin?

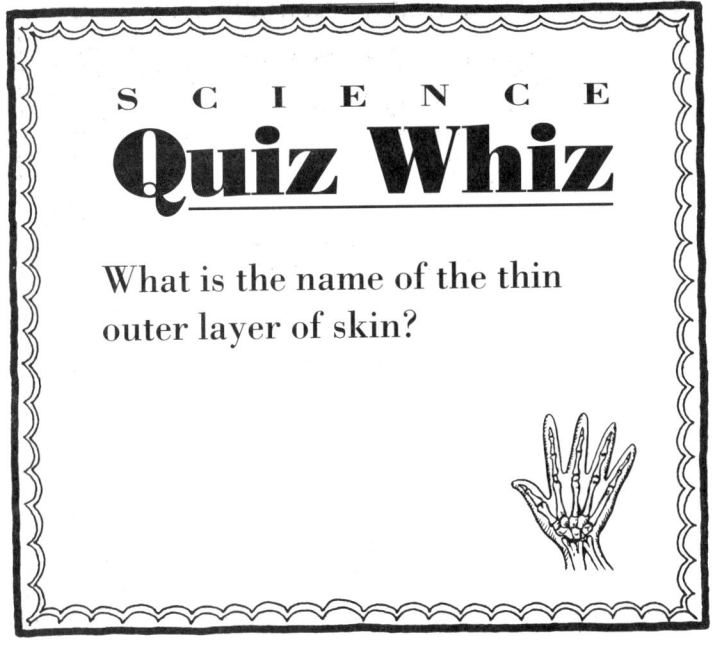

SCIENCE Quiz Whiz

Which of the following chemical elements is *not* normally found in the human body: carbon, hydrogen, silicon, or potassium?

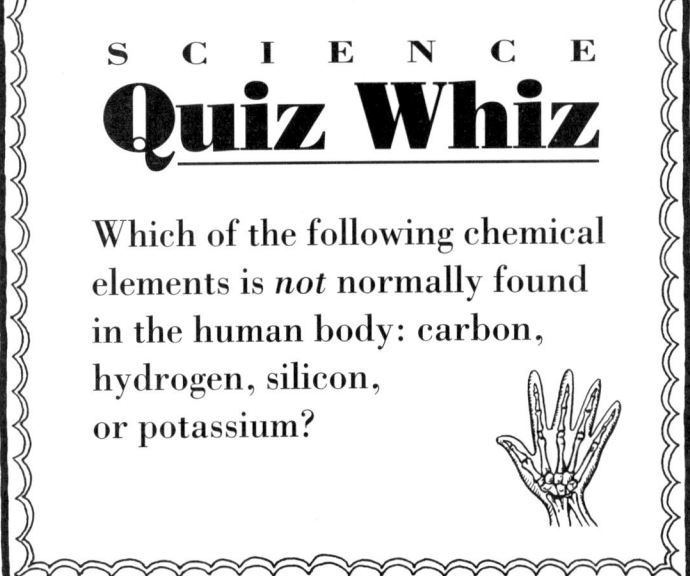

SCIENCE Quiz Whiz

What is the name of the soft tissue found inside bones which makes red blood cells?

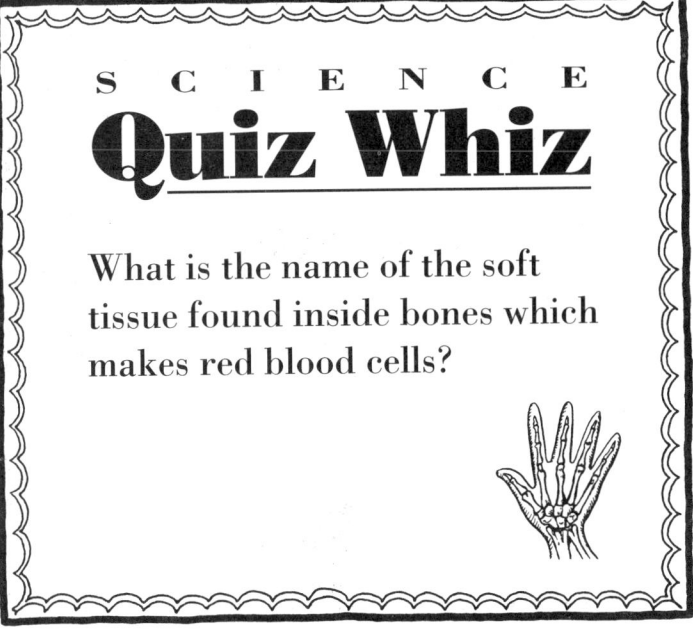

SCIENCE Quiz Whiz

Where in the human body is the "funny bone" located?

SCIENCE Quiz Whiz

How many chambers make up the human heart?

SCIENCE Quiz Whiz

Respiration is the process of taking in oxygen and expelling what gas?

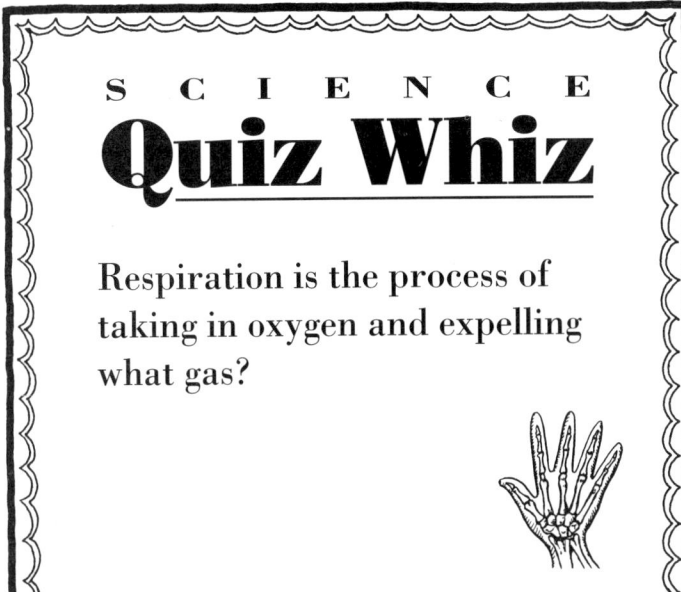

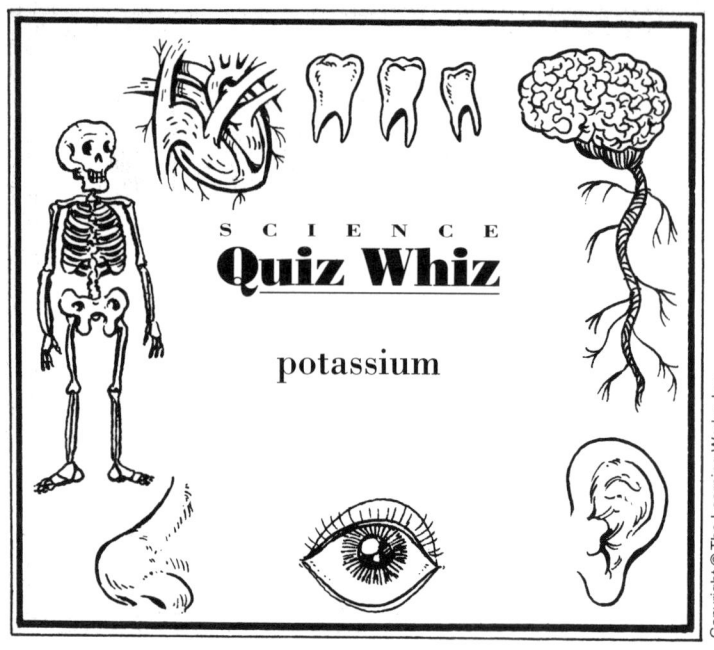

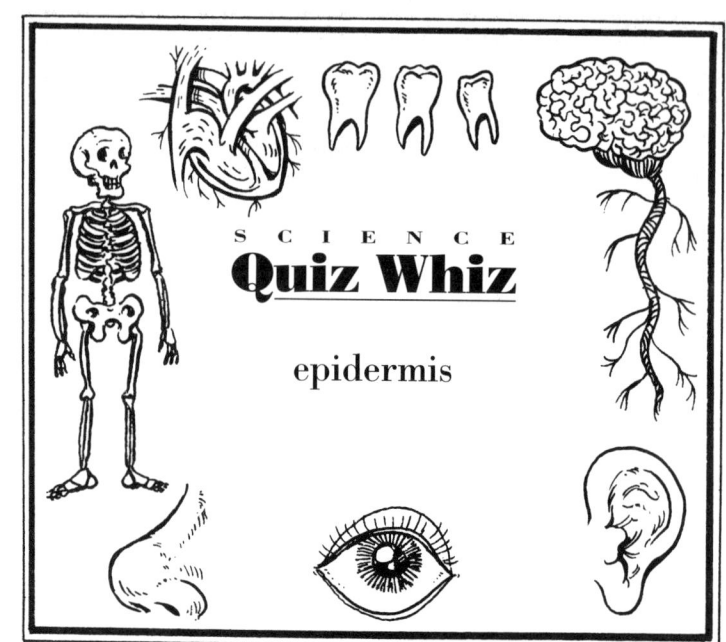

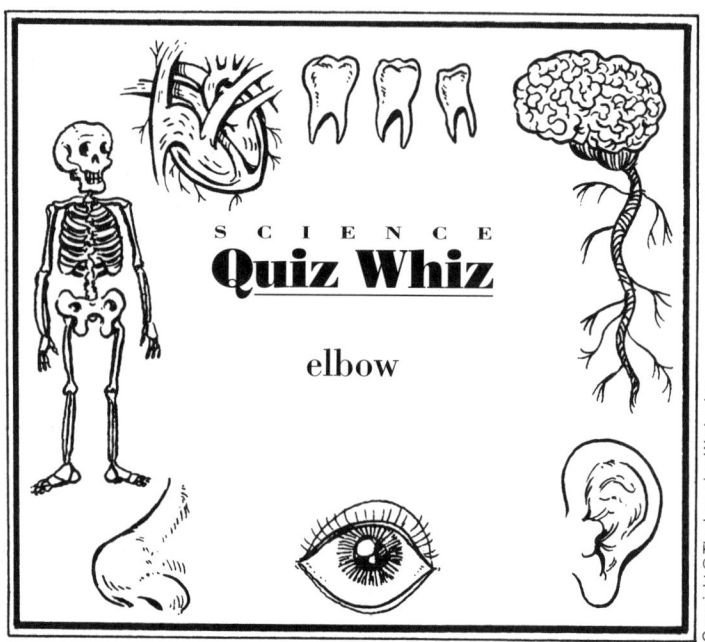

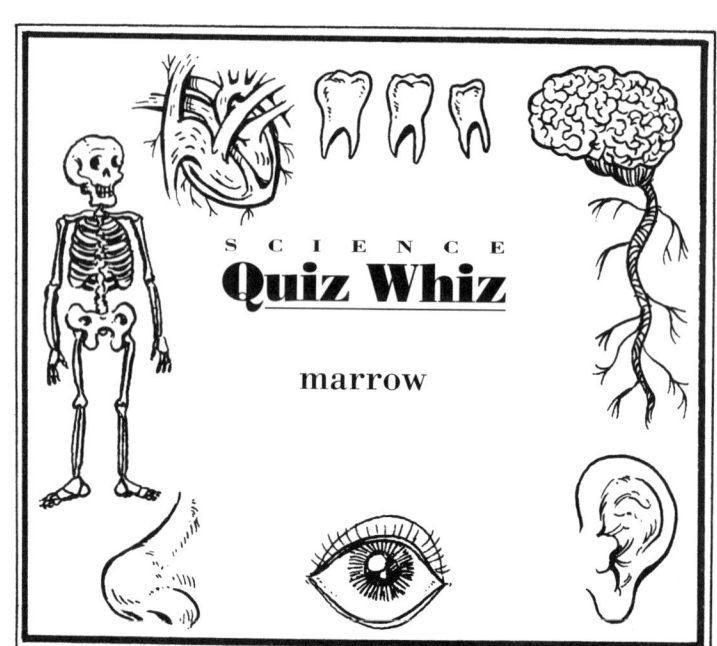

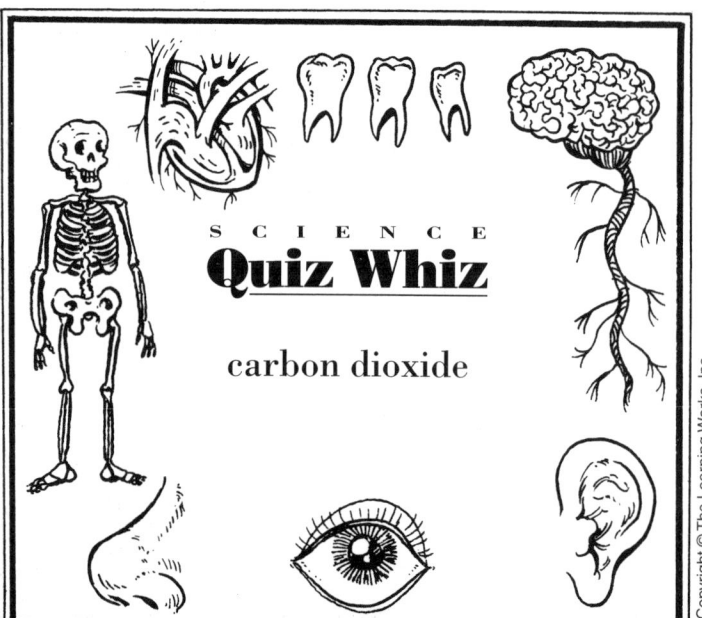

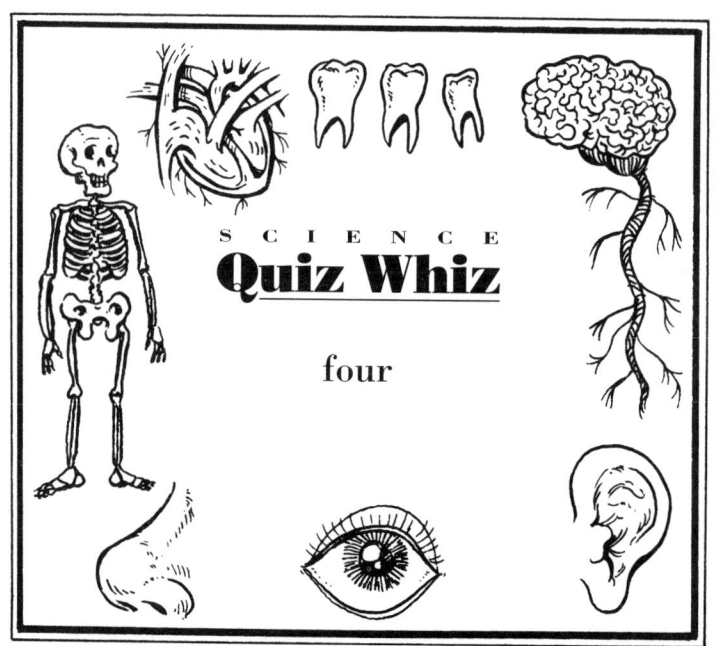

SCIENCE Quiz Whiz

What light-sensitive cells in the eye detect light and dark, but not color?

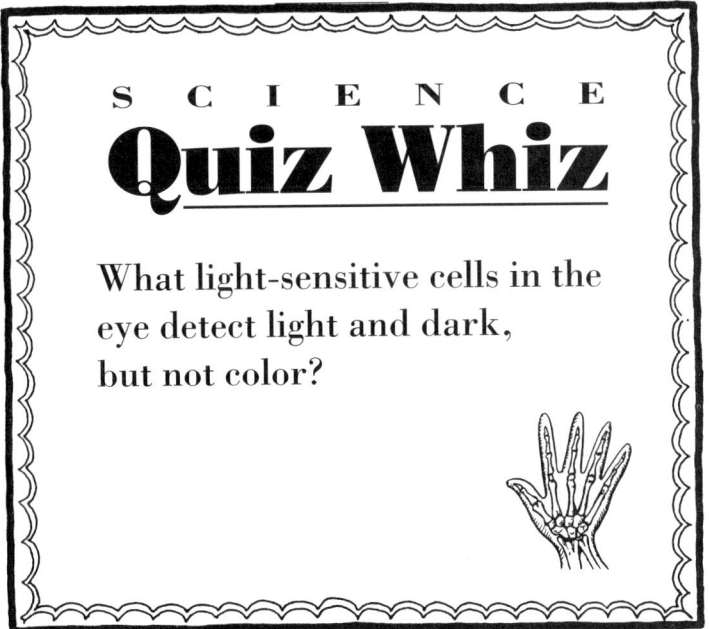

SCIENCE Quiz Whiz

What is the name of the muscular wall that separates the left and right sides of the heart?

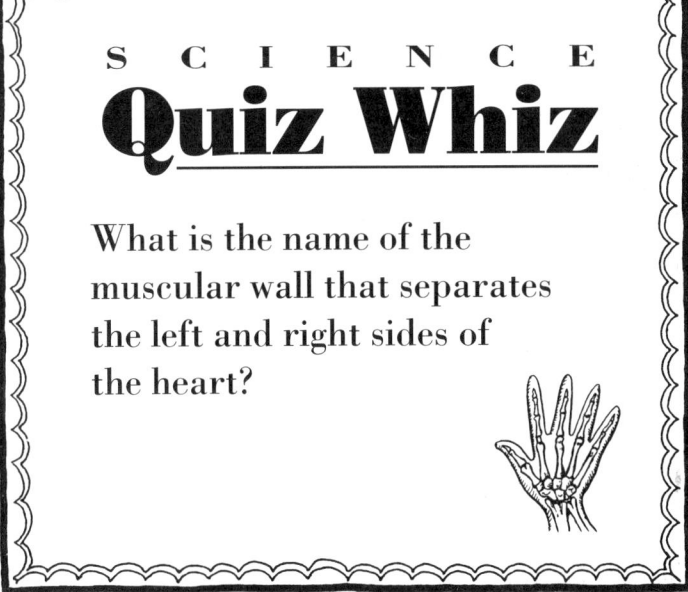

SCIENCE Quiz Whiz

What is the name of the dot-like opening in the center of the eye that allows light to enter the eye?

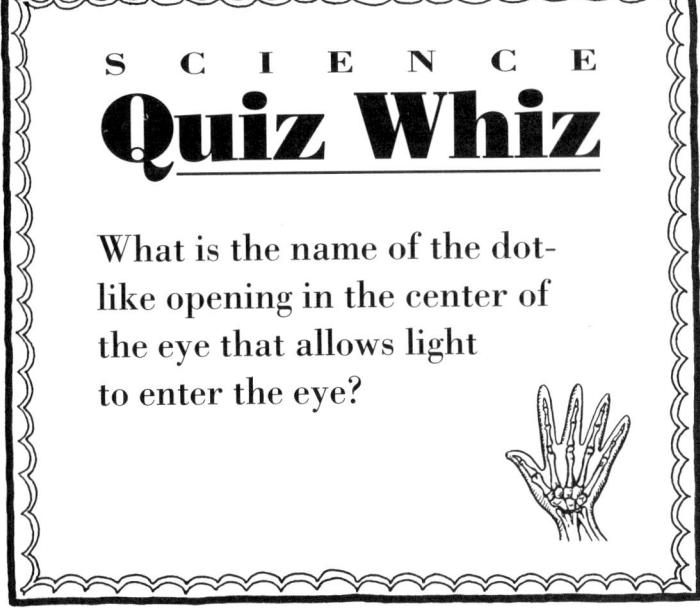

SCIENCE Quiz Whiz

What is each of the upper chambers of the heart called?

SCIENCE Quiz Whiz

The mouth, intestines, liver, and stomach are part of which system of the human body?

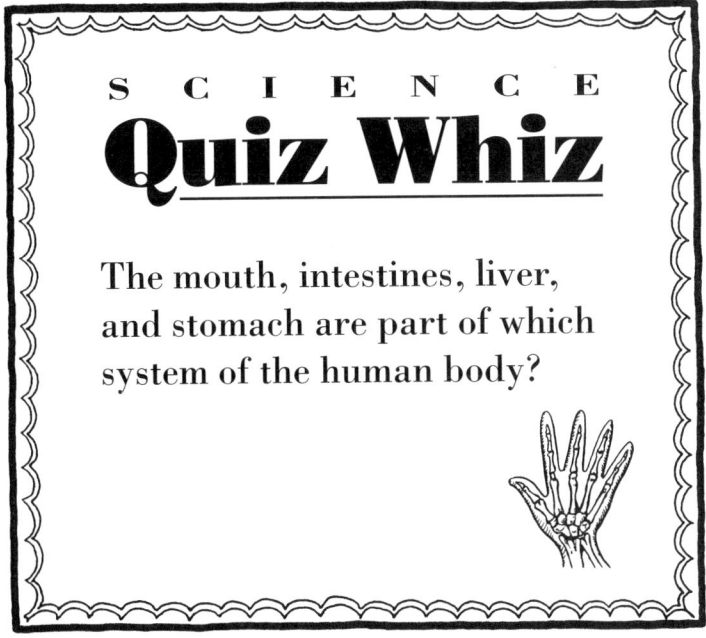

SCIENCE Quiz Whiz

What is the name of the tiny bundles of taste-sensitive nerves found on the surface of the tongue?

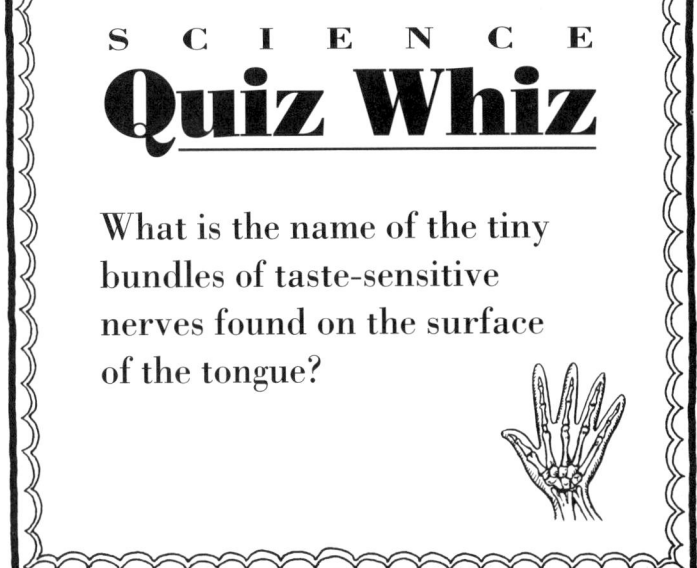

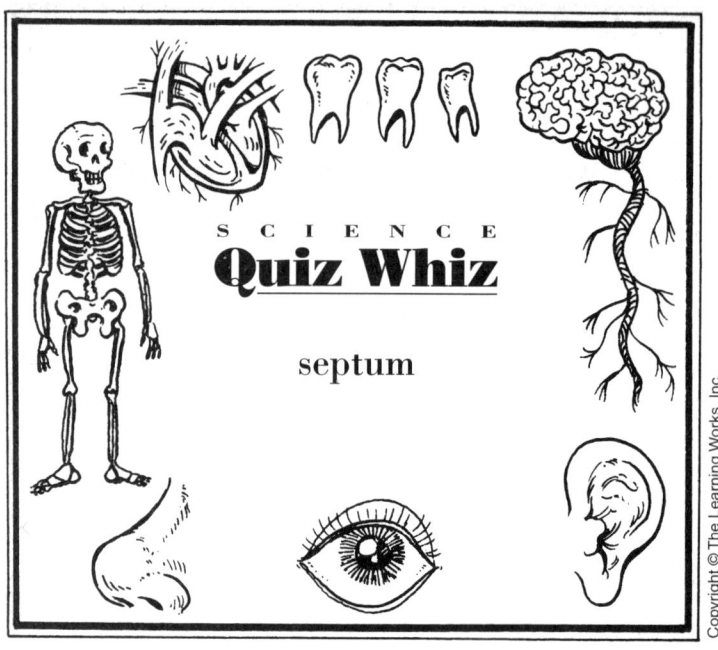
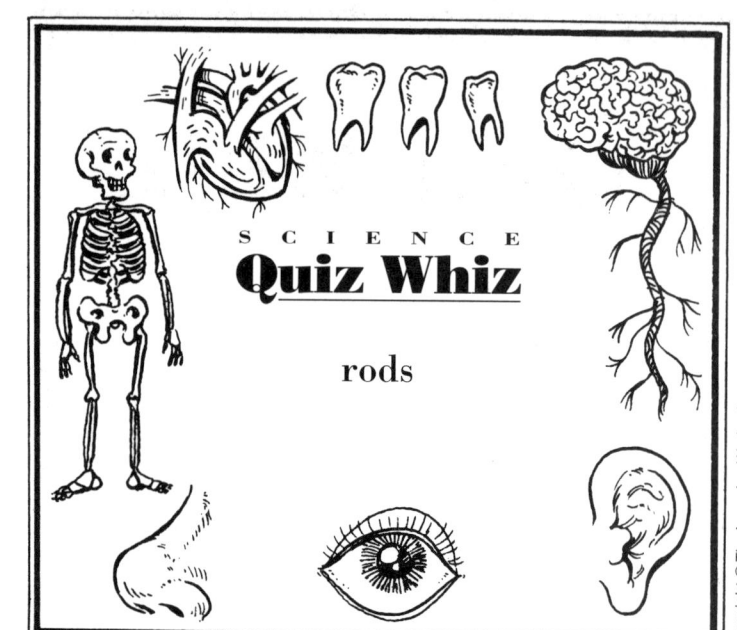
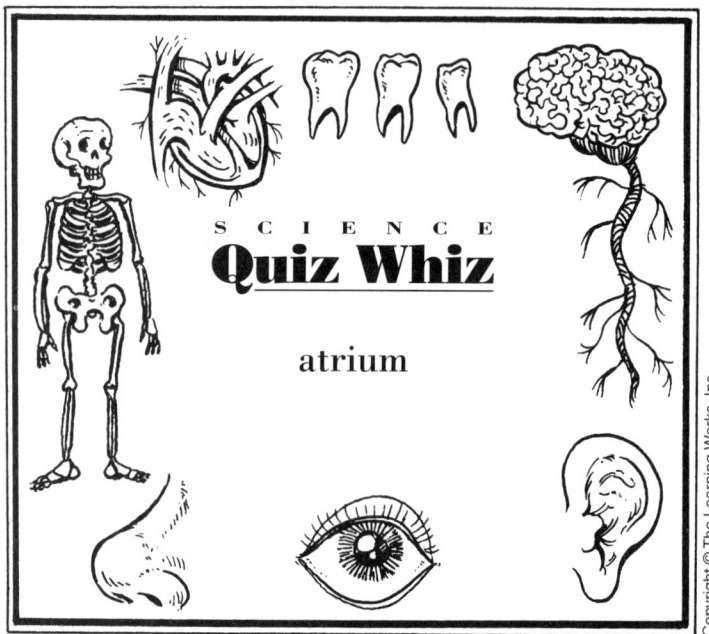
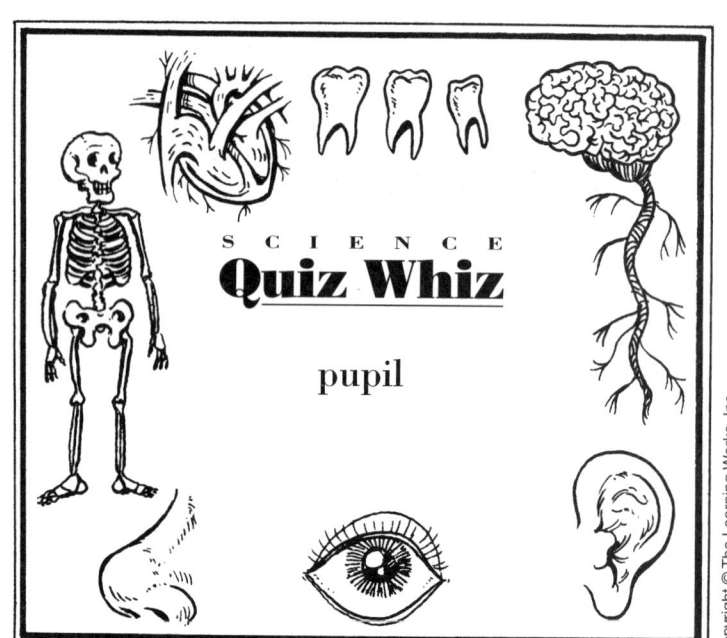
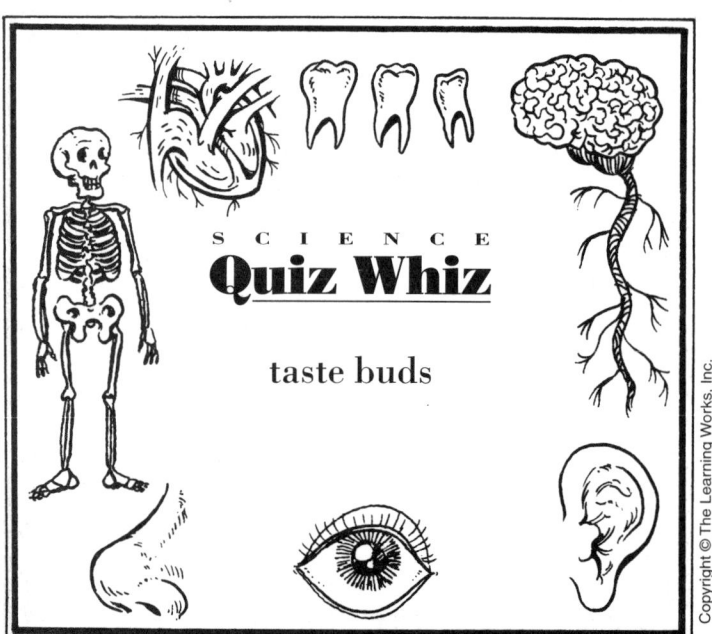
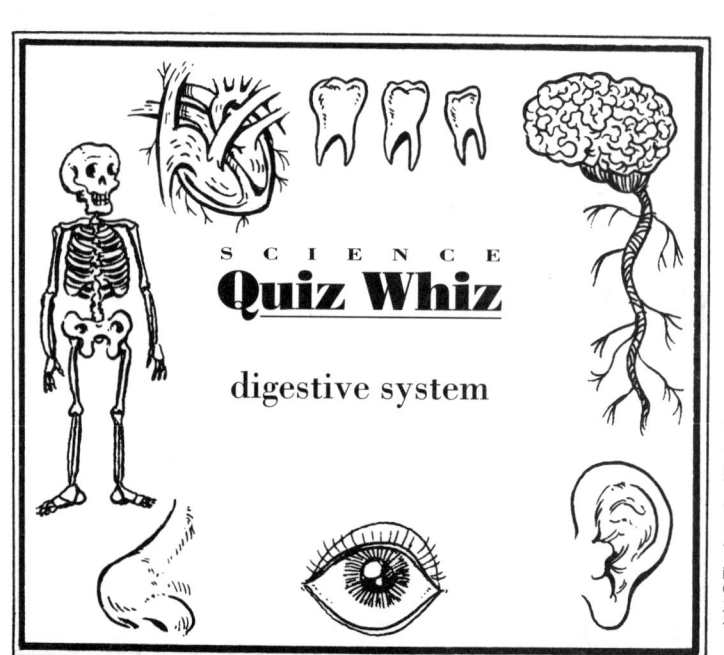

SCIENCE Quiz Whiz

What is another name for the kneecap?

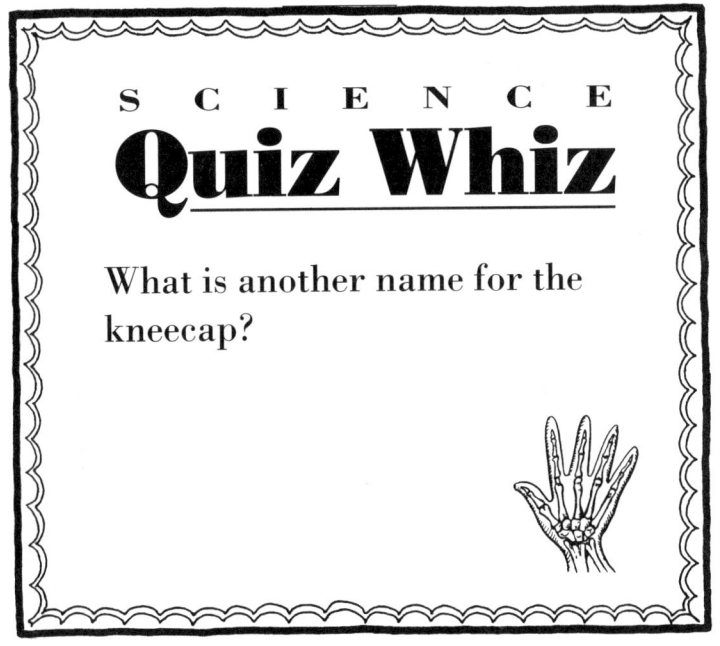

SCIENCE Quiz Whiz

What term is given to the chemical and mechanical breakdown of food in the body?

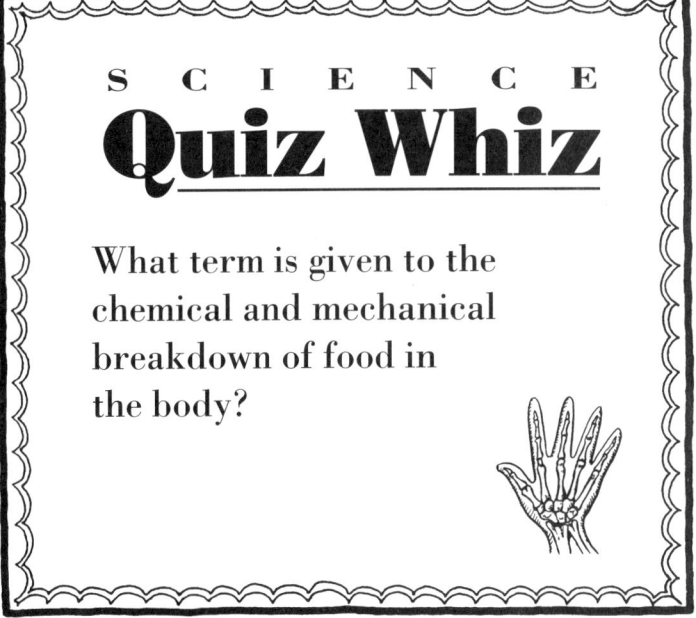

SCIENCE Quiz Whiz

Where is the smallest bone in the human body found?

SCIENCE Quiz Whiz

What is the name of the largest and most important part of the brain, where everything you learn is stored?

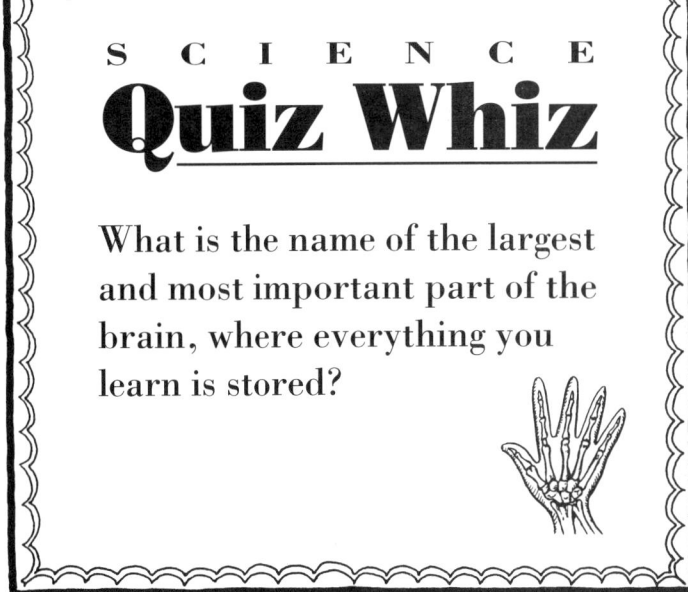

SCIENCE Quiz Whiz

Which bone is the largest and strongest in the human body?

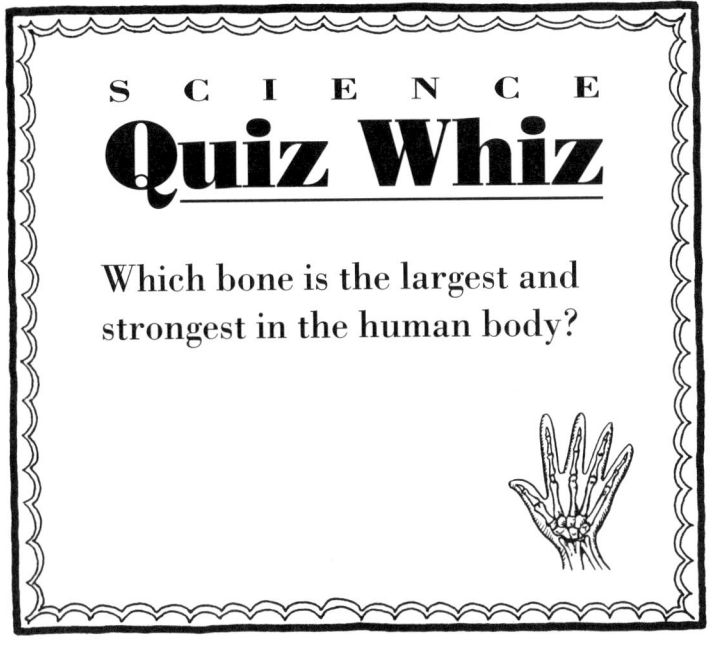

SCIENCE Quiz Whiz

What part of the brain, located at the base of the cerebrum, controls involuntary movements such as the beating of the heart?

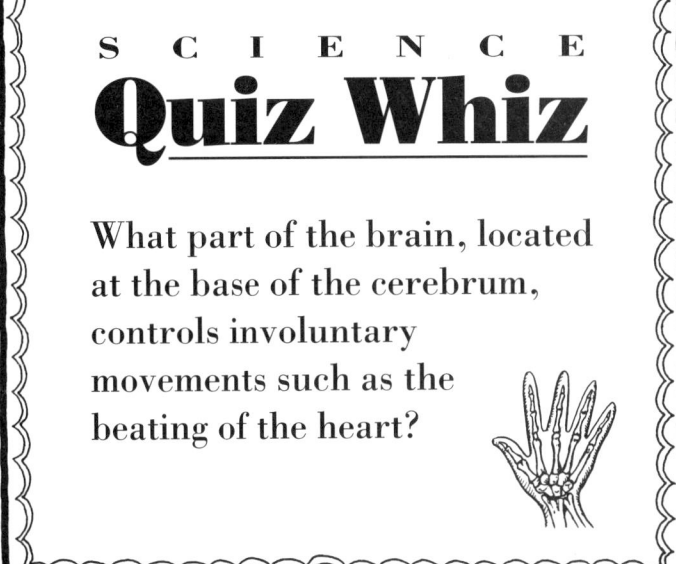

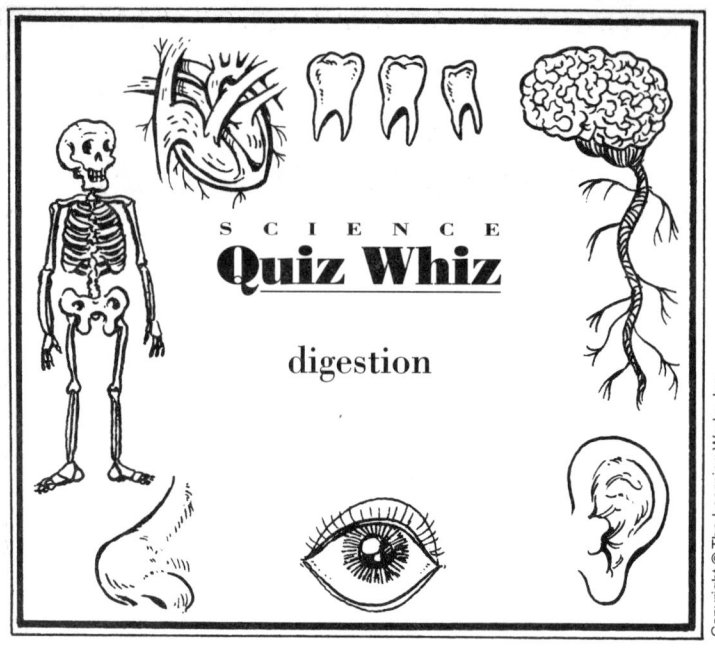

SCIENCE
Quiz Whiz

digestion

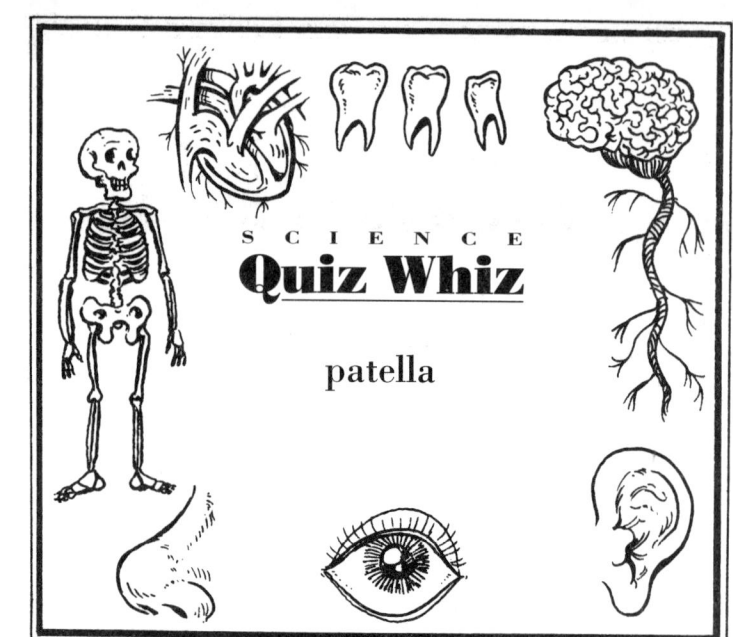

SCIENCE
Quiz Whiz

patella

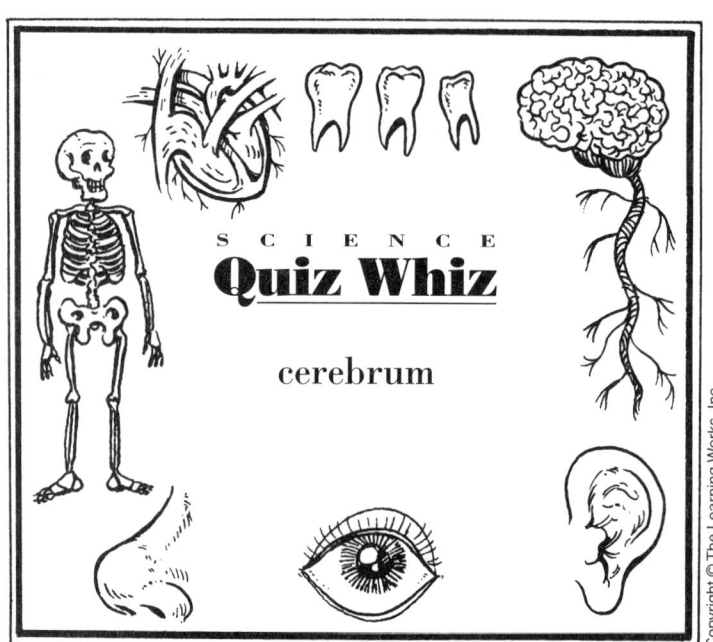

SCIENCE
Quiz Whiz

cerebrum

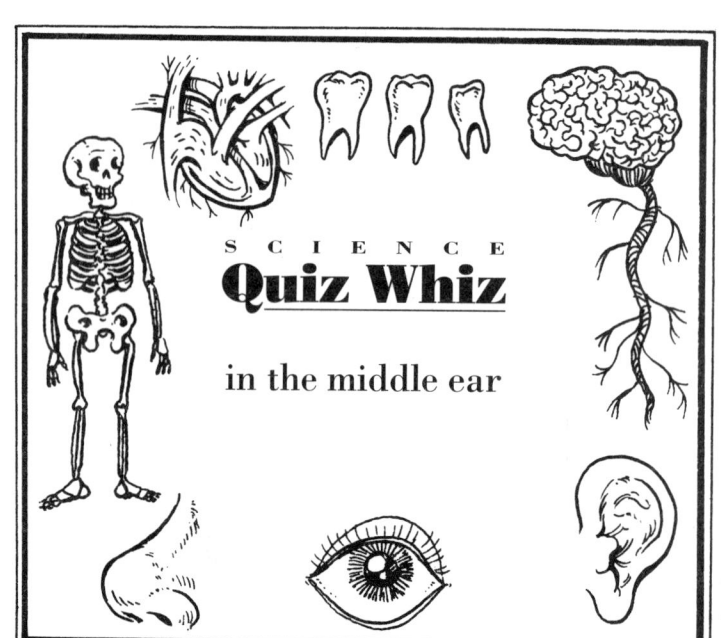

SCIENCE
Quiz Whiz

in the middle ear

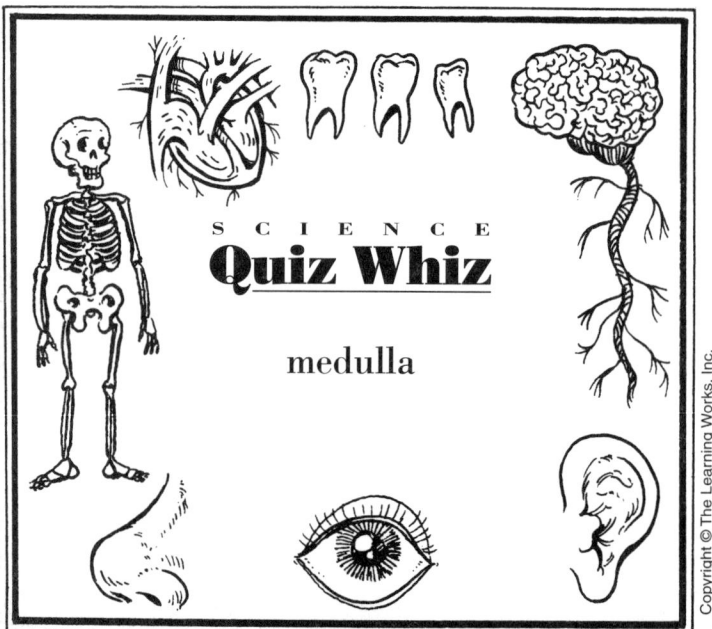

SCIENCE
Quiz Whiz

medulla

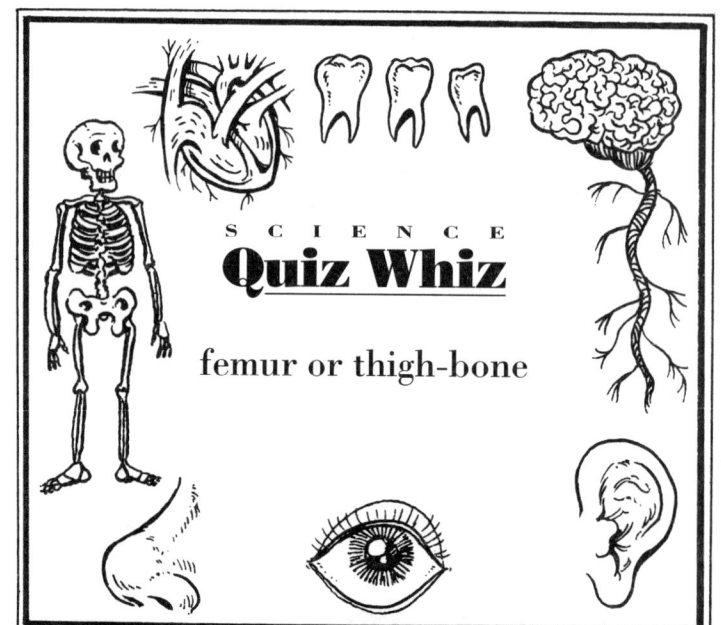

SCIENCE
Quiz Whiz

femur or thigh-bone

SCIENCE
Quiz Whiz

Who invented the telephone?

SCIENCE
Quiz Whiz

Which of the following did Charles Goodyear invent: steel plow, record player, microwave oven, or vulcanized rubber?

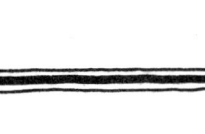

SCIENCE
Quiz Whiz

Which inventor is credited for the pasteurization of food products?

SCIENCE
Quiz Whiz

What mode of transportation did Igor Sikorsky invent?

SCIENCE
Quiz Whiz

Who invented the cotton gin?

SCIENCE
Quiz Whiz

Which British physicist discovered the existence of atomic nuclei?

SCIENCE Quiz Whiz

Who invented the lightning rod?

SCIENCE Quiz Whiz

Who invented the phonograph and the incandescent electric lamp?

SCIENCE Quiz Whiz

What was the name of the sheep that was the first animal made by cloning adult cells?

SCIENCE Quiz Whiz

George Washington Carver is credited for doing research on the industrial use of what food?

SCIENCE Quiz Whiz

What German physicist is famous for his theories of relativity and the equation $E=mc^2$?

SCIENCE Quiz Whiz

Who invented the first hand-held camera?

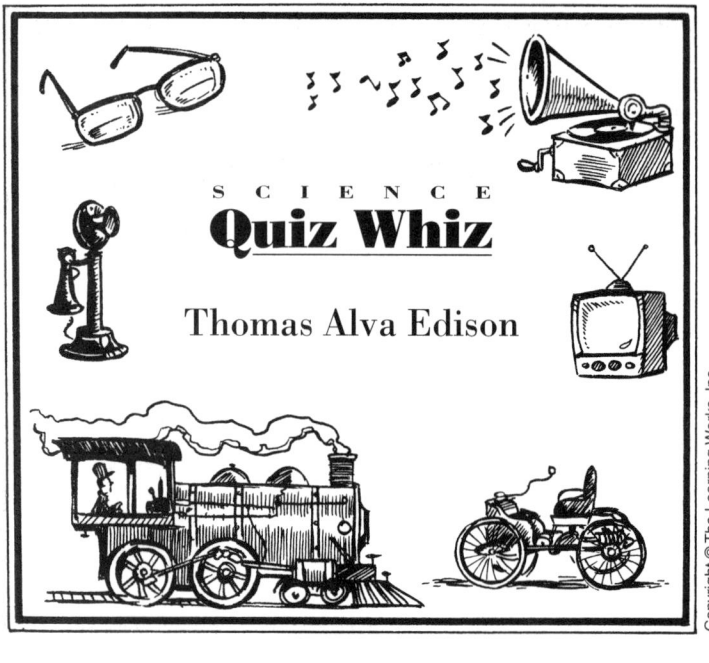

SCIENCE
Quiz Whiz

Charles Richter invented a scale to rate the intensity of what natural phenomena?

SCIENCE
Quiz Whiz

Who built the first nuclear reactor in the 1940s?

SCIENCE
Quiz Whiz

Which of the following did Philo Taylor Farnsworth invent: Teflon, nylon, television, or radio?

SCIENCE
Quiz Whiz

What did Guglielmo Marconi invent?

SCIENCE
Quiz Whiz

What French chemist, along with her husband, discovered the element radium?

SCIENCE
Quiz Whiz

Who invented the first automobile?

SCIENCE Quiz Whiz

Elisha Graves Otis is credited with what invention?

SCIENCE Quiz Whiz

Who invented a system of raised-point writing for literature and music to help the blind?

SCIENCE Quiz Whiz

What machine did Elias Howe invent?

SCIENCE Quiz Whiz

What kind of boat did Robert Fulton invent?

SCIENCE Quiz Whiz

What physicist launched the first liquid-fuel rocket in 1926?

SCIENCE Quiz Whiz

Who invented bifocal lenses?

SCIENCE
Quiz Whiz

What medicine did Alexander Fleming discover?

SCIENCE
Quiz Whiz

What office machine did Blaise Pascal invent?

SCIENCE
Quiz Whiz

Who was the first to discover a process of preserving food by freezing: Clarence Birdseye, Lee De Forest, Sara Lee, or Gottlieb Daimler?

SCIENCE
Quiz Whiz

Which of the following did Alfred Nobel invent: camera, dynamite, brakes, or electric lamp?

SCIENCE
Quiz Whiz

What American cartoonist invented animated cartoons?

SCIENCE
Quiz Whiz

Who invented the first revolver?

S C I E N C E
Quiz Whiz

Which Italian scientist proved that objects with different masses fall with the same velocity?

S C I E N C E
Quiz Whiz

Who invented the first American locomotive?

S C I E N C E
Quiz Whiz

Which of the following did Pierre Lallement invent: airplane, hydrofoil, bicycle, or tractor?

S C I E N C E
Quiz Whiz

Johannes Gutenberg developed which of the following: the power loom, printing from moveable type, the internal combustion engine, or the electric fan?

S C I E N C E
Quiz Whiz

What English scientist and mathematician discovered the principle of gravity?

S C I E N C E
Quiz Whiz

Who invented the magnetic telegraph?

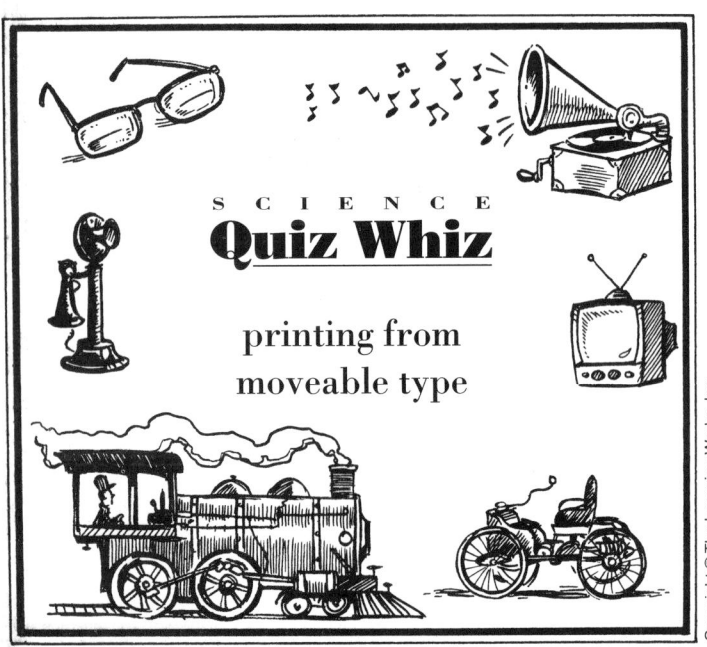

SCIENCE Quiz Whiz

Who was the first person to prove that lightning is a form of electricity?

SCIENCE Quiz Whiz

What is the name of the path through which an electric current flows?

SCIENCE Quiz Whiz

What is the name given to the ability to do work?

SCIENCE Quiz Whiz

What device has been used for hundreds of years to harness wind energy for such tasks as milling grain into flour?

SCIENCE Quiz Whiz

What is energy from the sun called?

SCIENCE Quiz Whiz

What is a machine called that changes energy of motion into electrical energy?

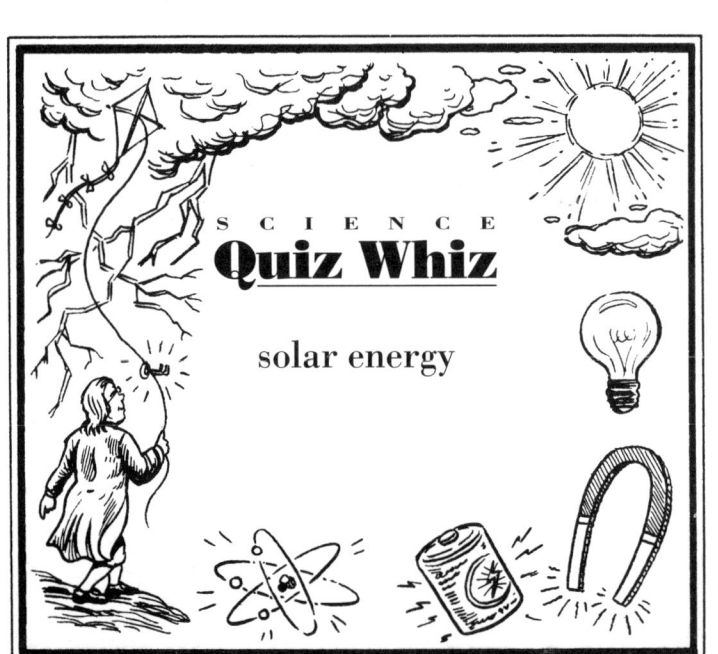

SCIENCE Quiz Whiz

What is the name given to anything that has mass and energy and takes up space?

SCIENCE Quiz Whiz

Which of the following is *not* a component of atoms: proton, neutron, crouton, electron?

SCIENCE Quiz Whiz

What kind of charge do the electrons in an atom have?

SCIENCE Quiz Whiz

Which of the following is a good conductor of electricity: rubber, glass, silver, or copper?

SCIENCE Quiz Whiz

What are kilowatt-hours used to measure?

SCIENCE Quiz Whiz

What is the thin coil of wire inside a light bulb called?

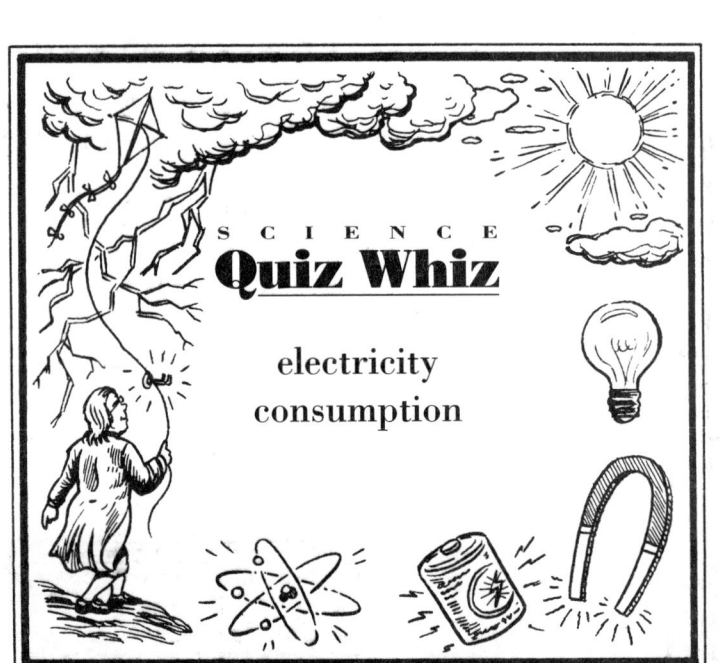

SCIENCE Quiz Whiz

What safety switch is used to break the flow of electricity when a circuit becomes overloaded?

SCIENCE Quiz Whiz

What is the term for an electric charge that does not move?

SCIENCE Quiz Whiz

From where does most of the energy on Earth come?

SCIENCE Quiz Whiz

What is the center of an atom called?

SCIENCE Quiz Whiz

What is oil called before it has been refined?

SCIENCE Quiz Whiz

Which is not a fossil fuel: coal, water, natural gas, or petroleum?

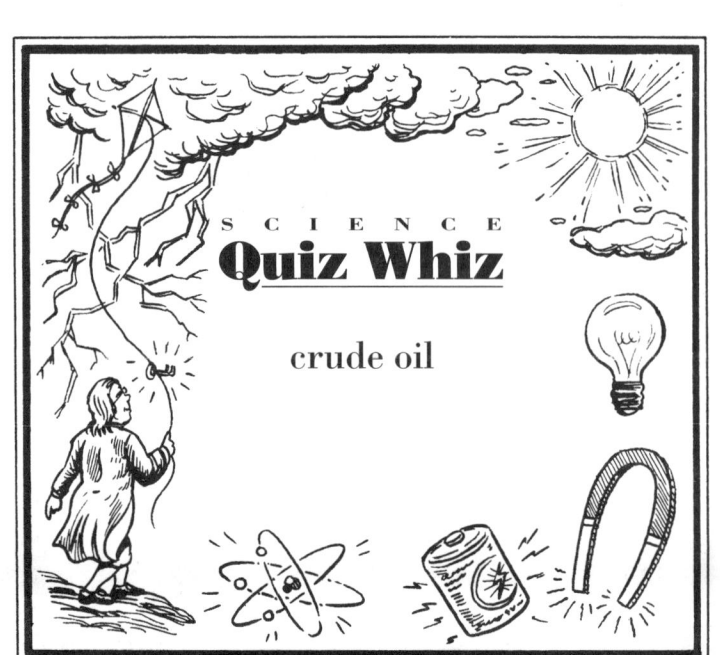

SCIENCE
Quiz Whiz

A kilowatt-hour is equal to how many watts of electricity used for one hour?

SCIENCE
Quiz Whiz

What is the name of the tiny, negatively charged particles that surround the nucleus of an atom?

SCIENCE
Quiz Whiz

What does an ammeter measure?

SCIENCE
Quiz Whiz

What is the device called that changes chemical energy to electrical energy?

SCIENCE
Quiz Whiz

What is the name for matter that is made up of only one kind of atom?

SCIENCE
Quiz Whiz

Which of the following is *not* an electrical insulator: water, plastic, wood, or rubber?

electrons

1,000 watts

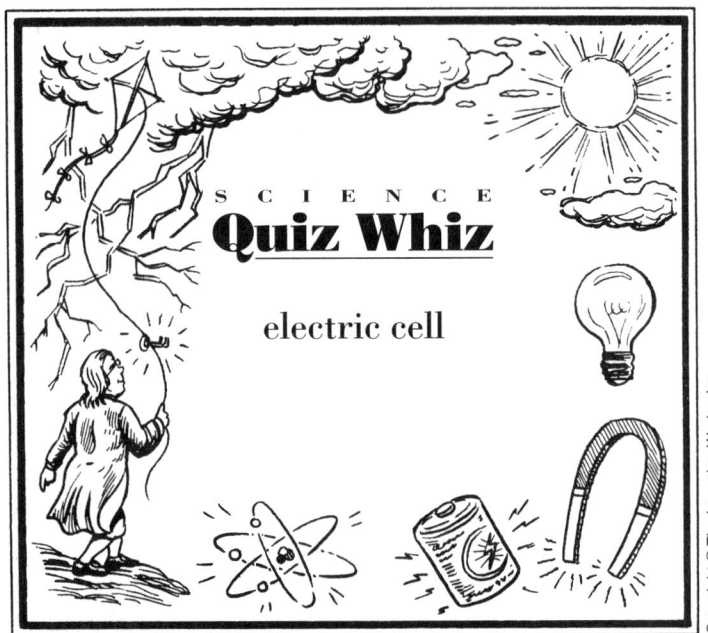
electric cell

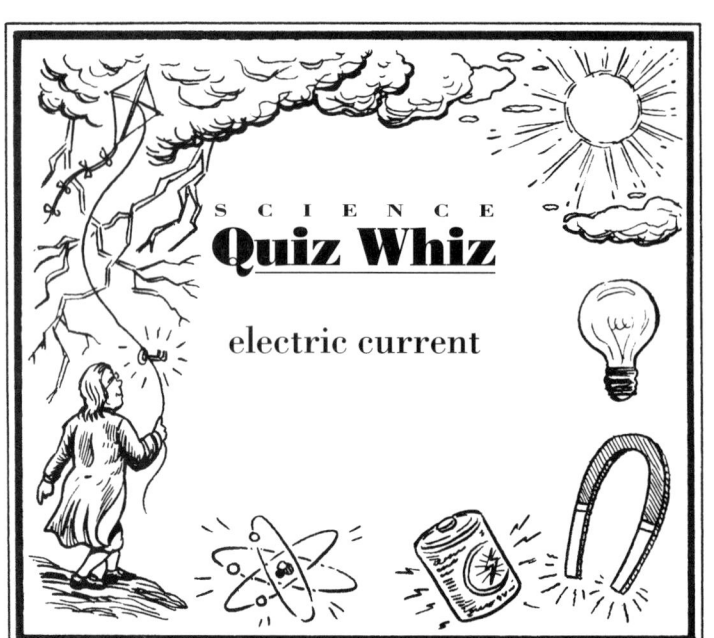
electric current

water

element

SCIENCE Quiz Whiz

What is the term for a substance through which an electric current cannot easily flow?

SCIENCE Quiz Whiz

What is the instrument called that can change electrical voltage?

SCIENCE Quiz Whiz

What electrical safety device has a piece of metal that melts when a current becomes too strong?

SCIENCE Quiz Whiz

What metal is used to make the filaments of light bulbs?

SCIENCE Quiz Whiz

What is the term for energy trapped in hot rocks below the earth's surface?

SCIENCE Quiz Whiz

What is the unit called "hertz," abbreviated "Hz," used to measure?

SCIENCE Quiz Whiz

What is a material called that allows an electric current to flow through it?

SCIENCE Quiz Whiz

What type of electric current flows in one direction only?

SCIENCE Quiz Whiz

What device consists of a core of iron surrounded by a coil of wire?

SCIENCE Quiz Whiz

What type of electric current changes direction repeatedly?

SCIENCE Quiz Whiz

What unit is used to measure the force of an electric current?

SCIENCE Quiz Whiz

What are the names of the two poles of a battery?

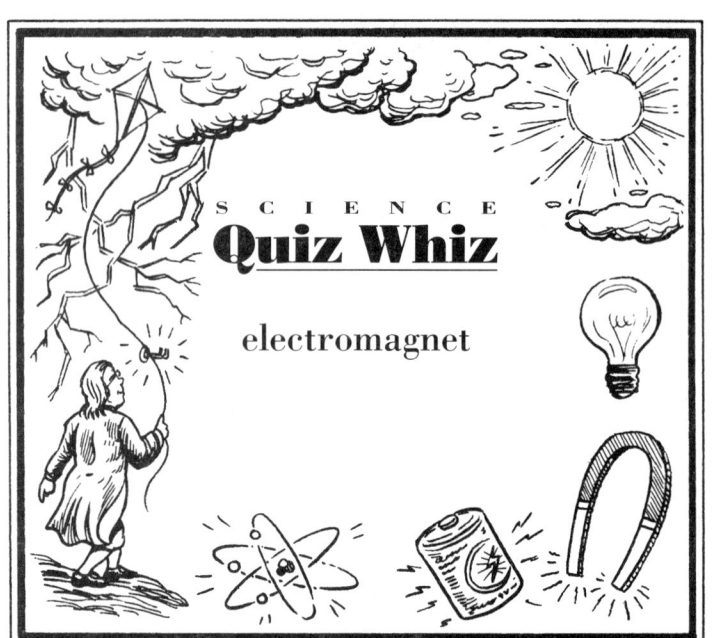

SCIENCE Quiz Whiz

What instrument is used to measure air pressure?

SCIENCE Quiz Whiz

What two elements make up water?

SCIENCE Quiz Whiz

In the United States, what scale is used to measure air temperature?

SCIENCE Quiz Whiz

What term is given to the amount of moisture in the air?

SCIENCE Quiz Whiz

What are the three forms water can take?

SCIENCE Quiz Whiz

What is a person called who reports and forecasts the weather?

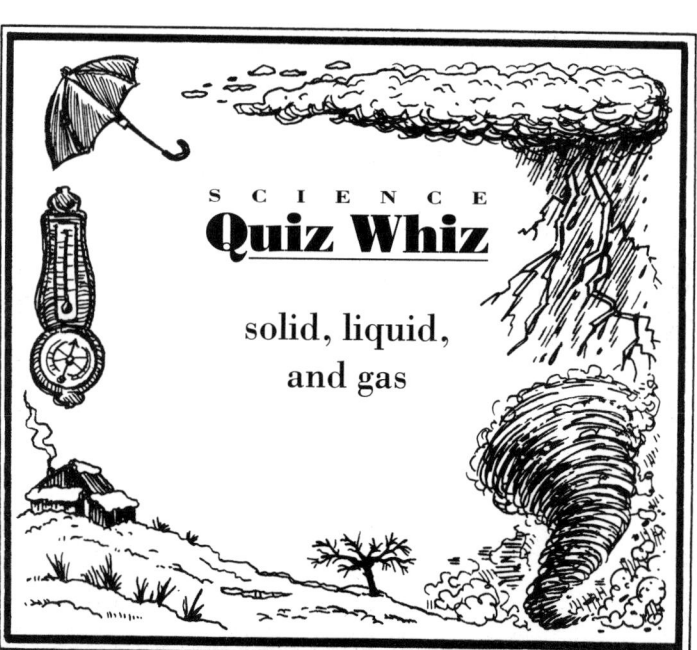

SCIENCE Quiz Whiz

Which state has the greatest number of tornadoes each year?

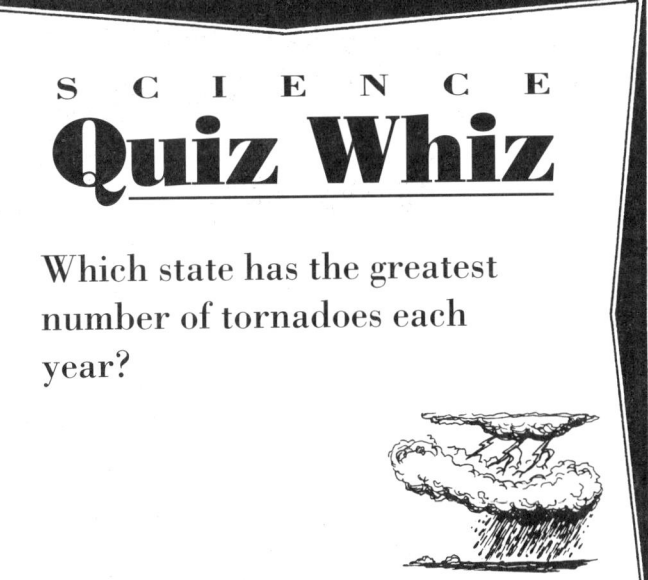

SCIENCE Quiz Whiz

What is the term for the circulation of water from the land to the atmosphere and back to land?

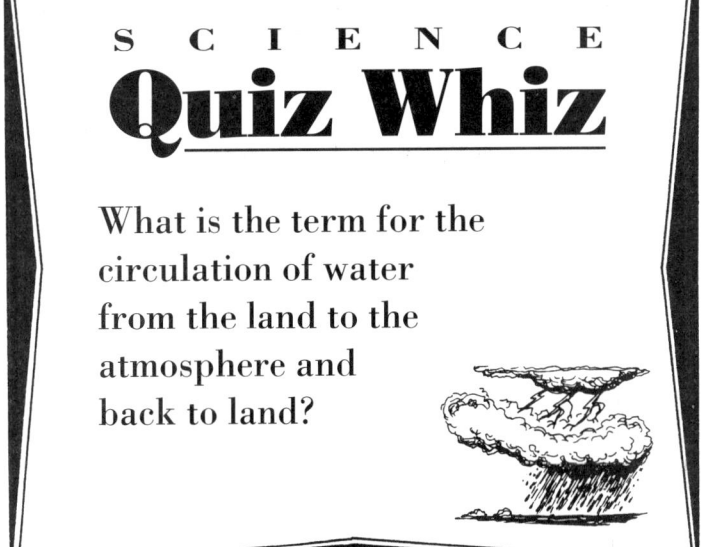

SCIENCE Quiz Whiz

What is the chemical symbol for water?

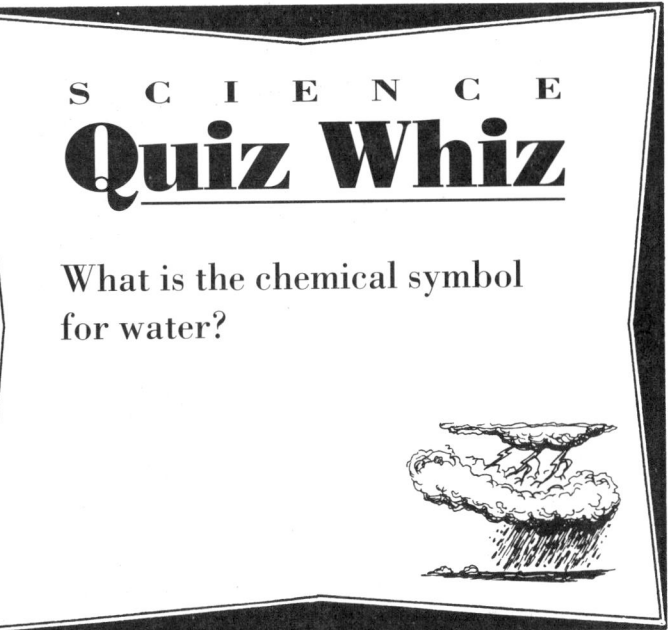

SCIENCE Quiz Whiz

Water covers approximately what percentage of Earth's surface: 25, 45, 60, or 70?

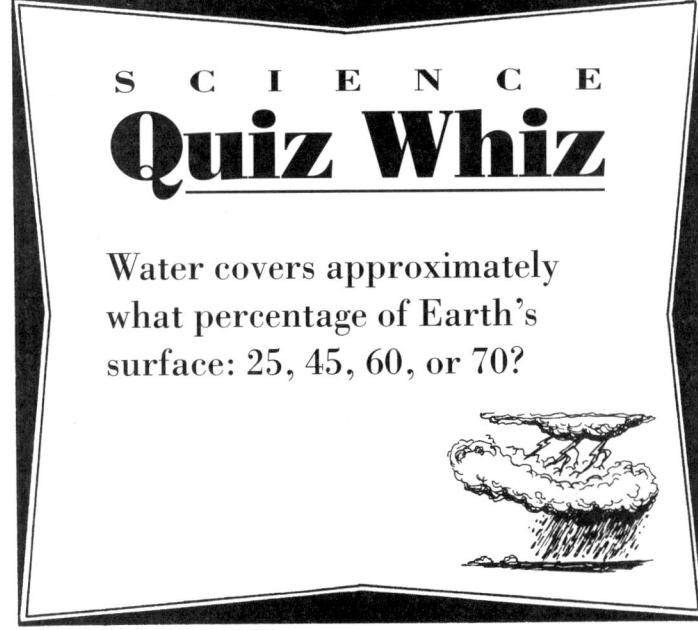

SCIENCE Quiz Whiz

What are winds that blow at more than 74 miles per hour called?

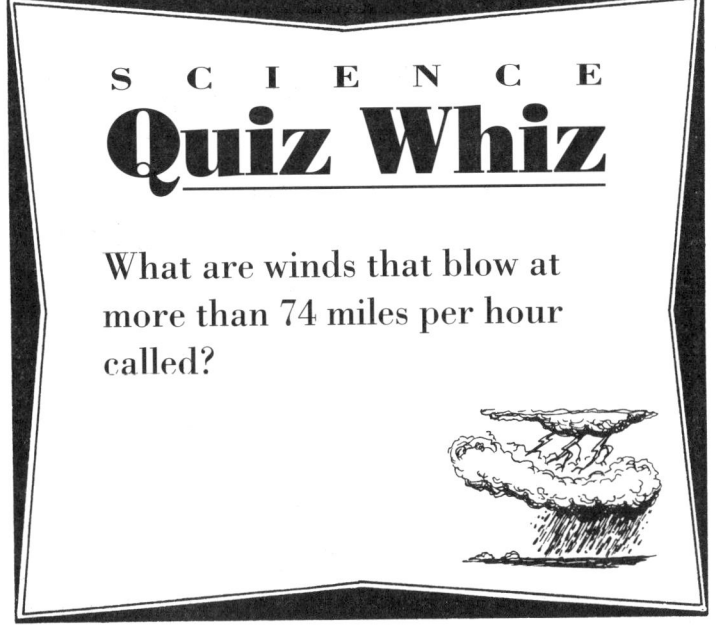

SCIENCE Quiz Whiz

What is the process called by which a liquid turns into a gas?

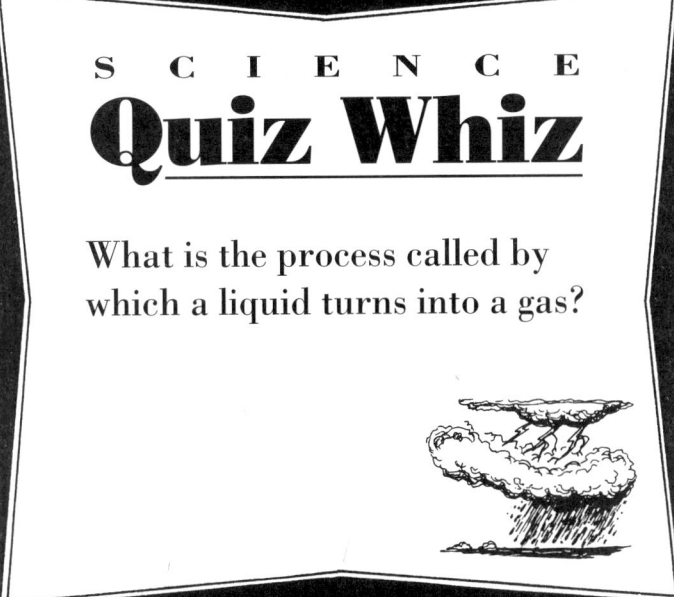

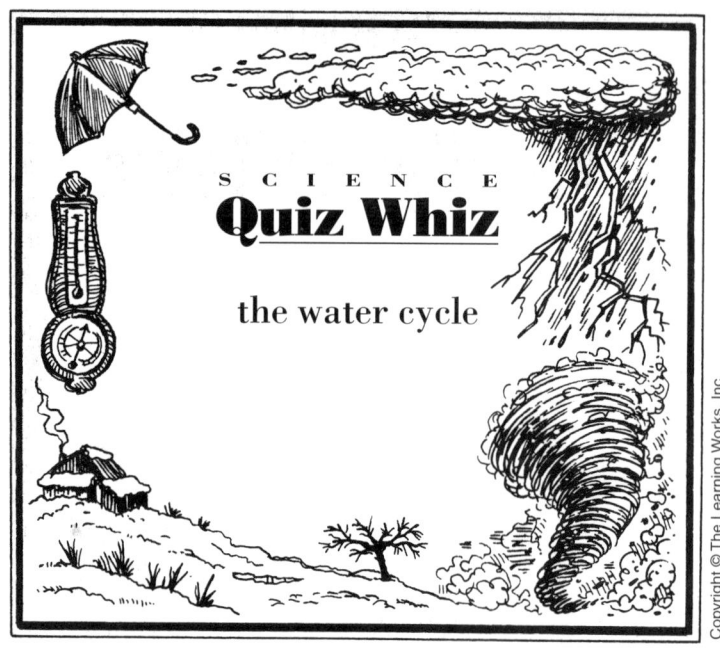

SCIENCE
Quiz Whiz
the water cycle

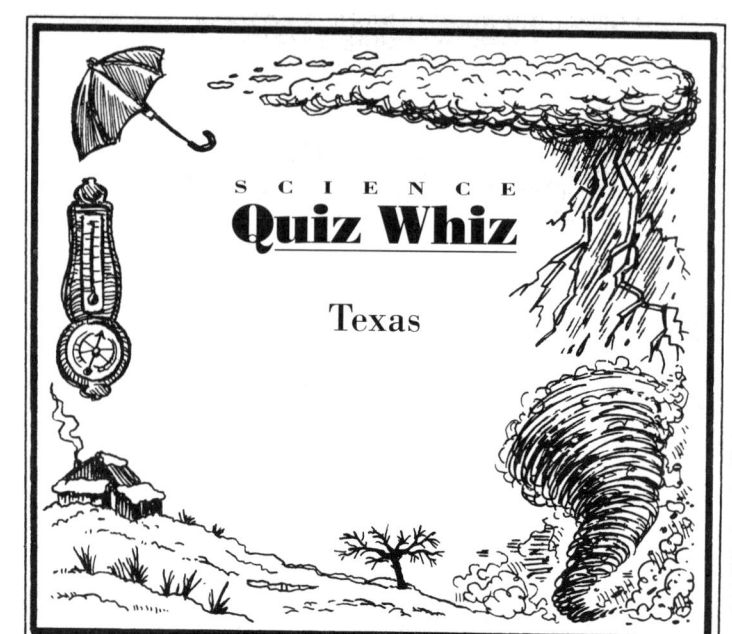

SCIENCE
Quiz Whiz
Texas

SCIENCE
Quiz Whiz
70 percent

SCIENCE
Quiz Whiz
H_2O

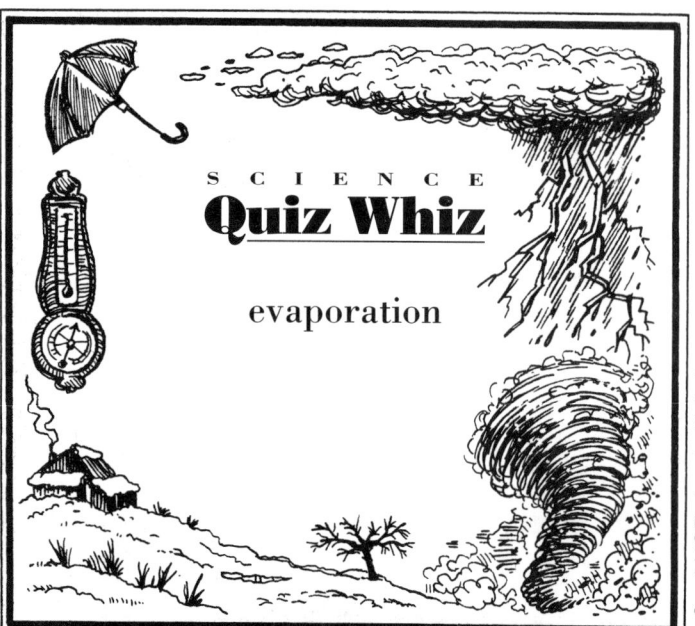

SCIENCE
Quiz Whiz
evaporation

SCIENCE
Quiz Whiz
hurricanes

SCIENCE Quiz Whiz

During a storm, what is the shock wave called that comes from the heating and cooling of air near lightning?

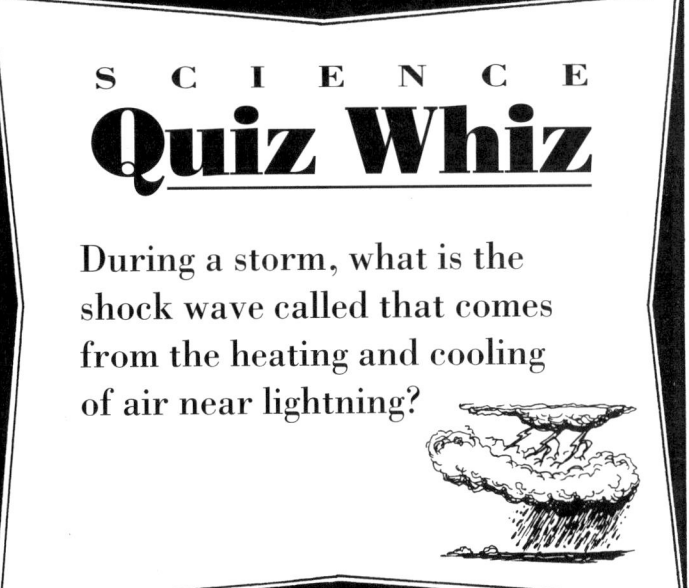

SCIENCE Quiz Whiz

What term is given to water that is a gas?

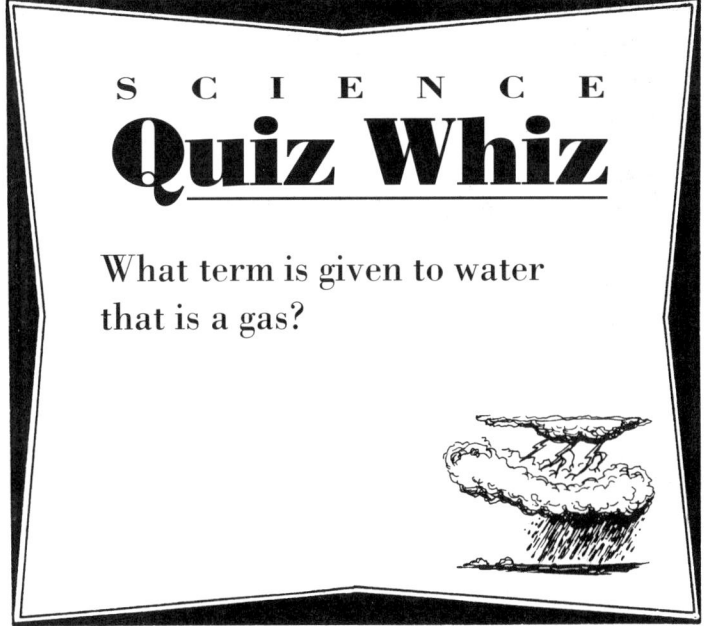

SCIENCE Quiz Whiz

Which type of cloud is most likely to produce rain: cumulus, stratus, nebulous, or cumulo-nimbus?

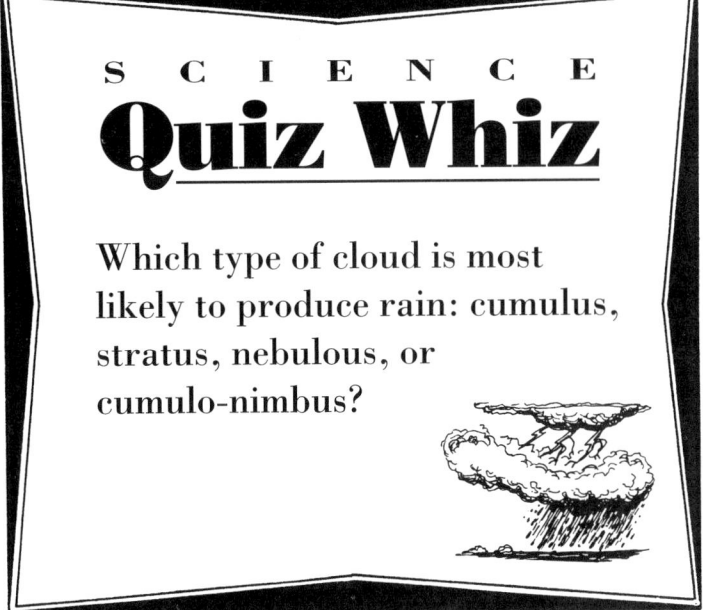

SCIENCE Quiz Whiz

What is the scientific word for rain, sleet, hail, or snow?

SCIENCE Quiz Whiz

In condensation, a gas changes into what form?

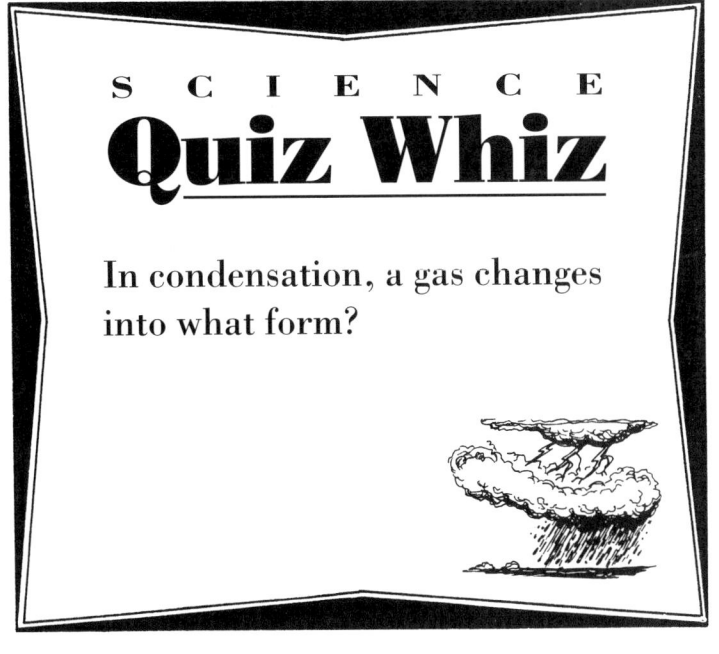

SCIENCE Quiz Whiz

What term is given to water that is solid?

SCIENCE Quiz Whiz

What is the freezing point of water at sea level?

SCIENCE Quiz Whiz

What element is pumped into drinking water in treatment plants to kill harmful bacteria?

SCIENCE Quiz Whiz

What is the term for the wearing away of soil and rock by water, ice, and wind?

SCIENCE Quiz Whiz

What term is given to the process of bringing water to dry land?

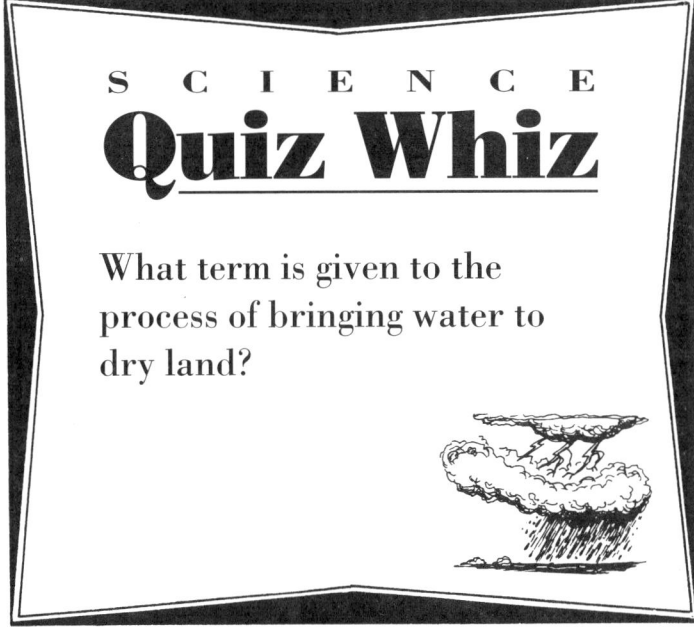

SCIENCE Quiz Whiz

A tsunami is a type of: cloud, wave, high temperature, or strong wind?

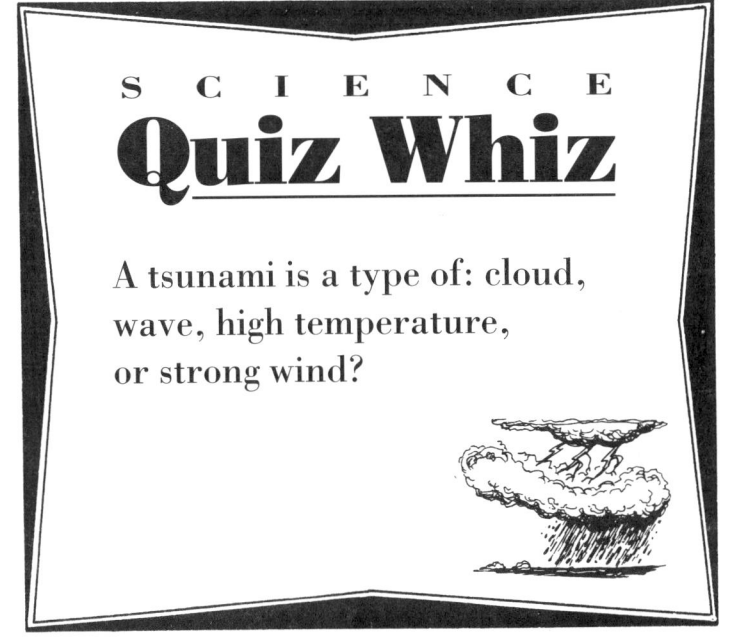

SCIENCE Quiz Whiz

What instrument is used to measure the amount of moisture in the air?

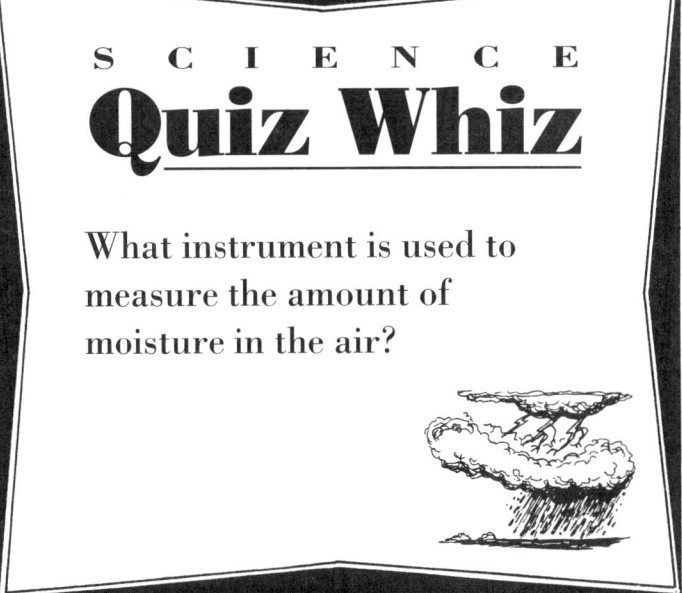

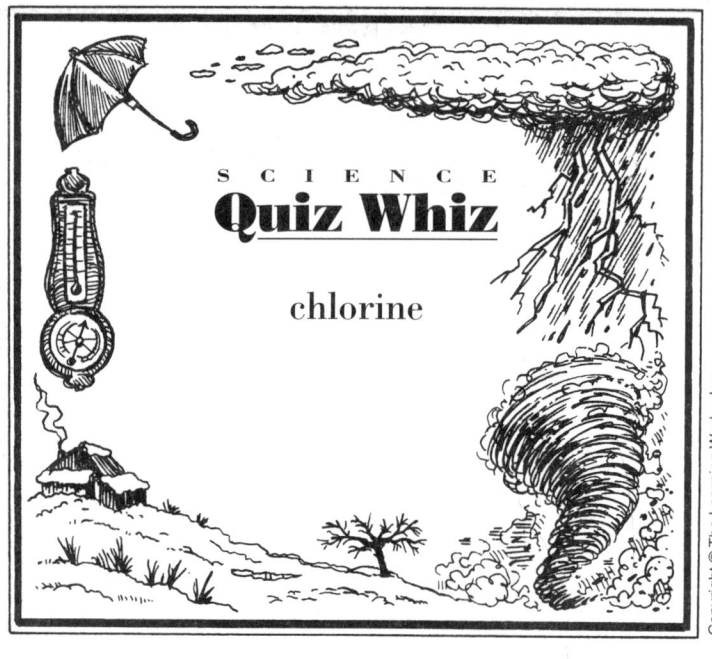

SCIENCE Quiz Whiz

What atmospheric effect is produced by sunlight shining through rain?

SCIENCE Quiz Whiz

In which direction does weather move across the United States?

SCIENCE Quiz Whiz

Which direction do hurricanes spin, clockwise or counter-clockwise?

SCIENCE Quiz Whiz

During a storm, why do we see lightning before we hear thunder?

SCIENCE Quiz Whiz

What is the calm center of a hurricane called?

SCIENCE Quiz Whiz

What is the process called by which salt is taken out of sea water?

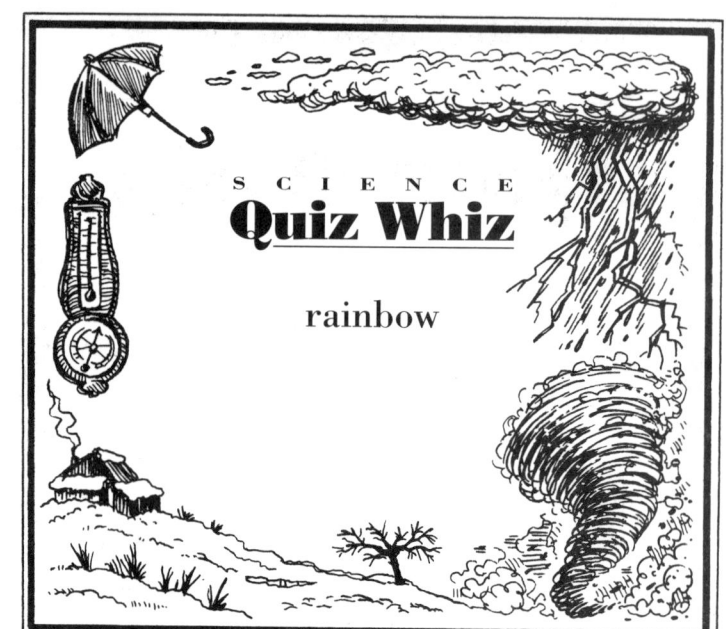

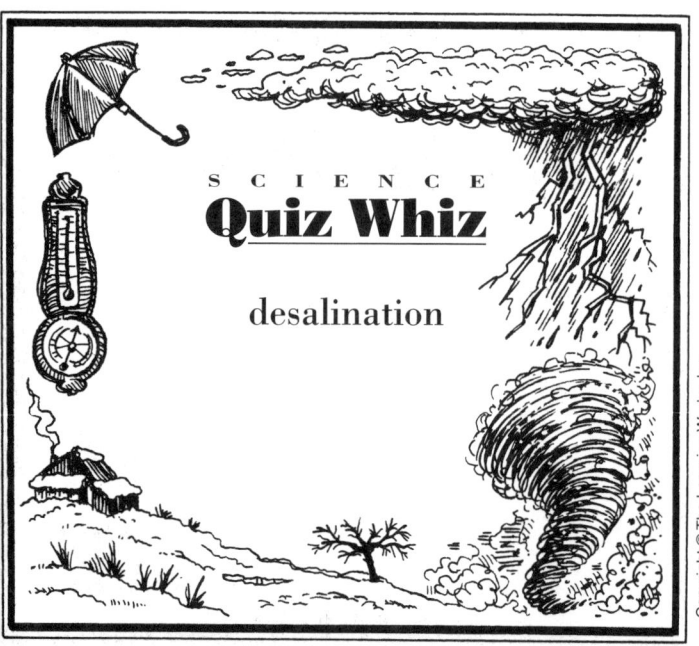

SCIENCE
Quiz Whiz

What instrument is used to measure wind speed?

SCIENCE
Quiz Whiz

About how many quarts of water does the human body need each day?

SCIENCE
Quiz Whiz

What is the process by which water in the atmosphere is deposited as droplets on surfaces, such as morning dew on grass?

SCIENCE
Quiz Whiz

What name is given to a place that stores water for future use: reservoir, aqueduct, kiln, or delta?

SCIENCE
Quiz Whiz

What kind of cloud is shown in the illustration on this card?

SCIENCE
Quiz Whiz

What is the name of the heavy seasonal rains that drench India every spring?

two quarts

anemometer

reservoir

condensation

monsoon

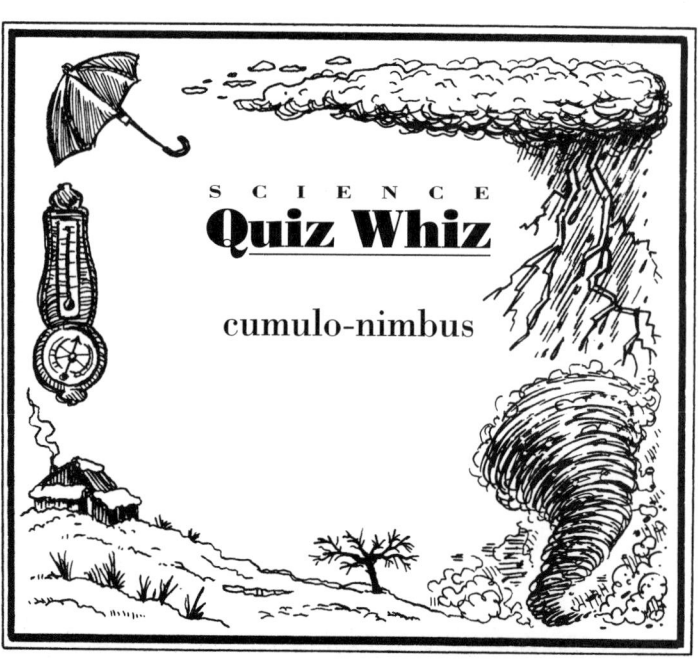
cumulo-nimbus

SCIENCE Quiz Whiz

What is the name given to a group of stars bound together by gravity?

SCIENCE Quiz Whiz

What is the study of celestial bodies called?

SCIENCE Quiz Whiz

What kind of eclipse occurs when sunlight on its way to the moon is blocked by Earth?

SCIENCE Quiz Whiz

What is the term for a cloud of dust and gas in space?

SCIENCE Quiz Whiz

Which planet in our solar system has the most moons?

SCIENCE Quiz Whiz

Which is the smallest planet in our solar system?

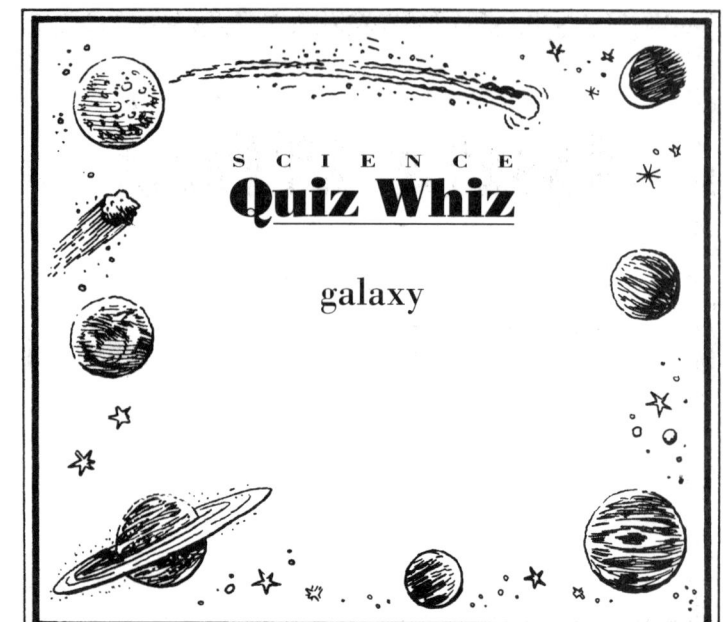

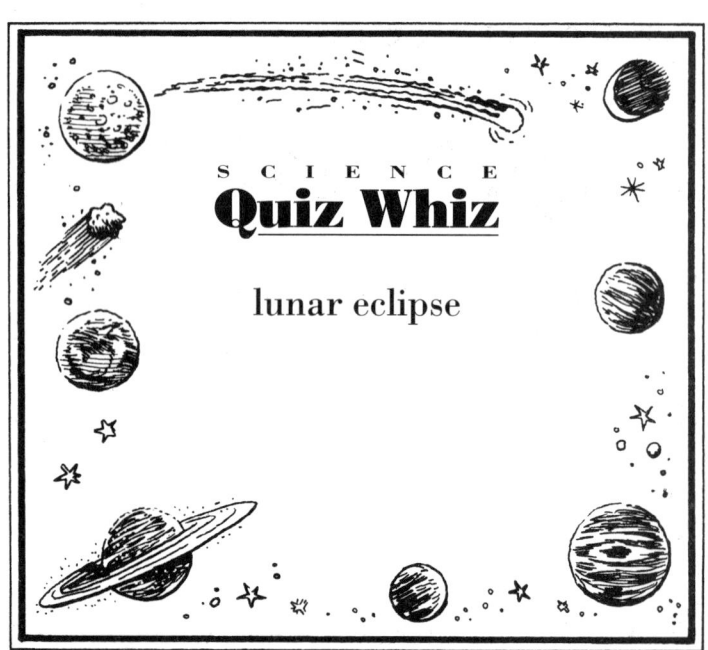

SCIENCE
Quiz Whiz

What term is applied to stars that explode?

SCIENCE
Quiz Whiz

Which planet is closest to the sun?

SCIENCE
Quiz Whiz

In which galaxy is Earth located?

SCIENCE
Quiz Whiz

Which planet is closest to Earth?

SCIENCE
Quiz Whiz

What celestial objects are so dense that not even light can escape their gravitation?

SCIENCE
Quiz Whiz

What term is used to describe a star's brightness?

Mercury

nova

Venus

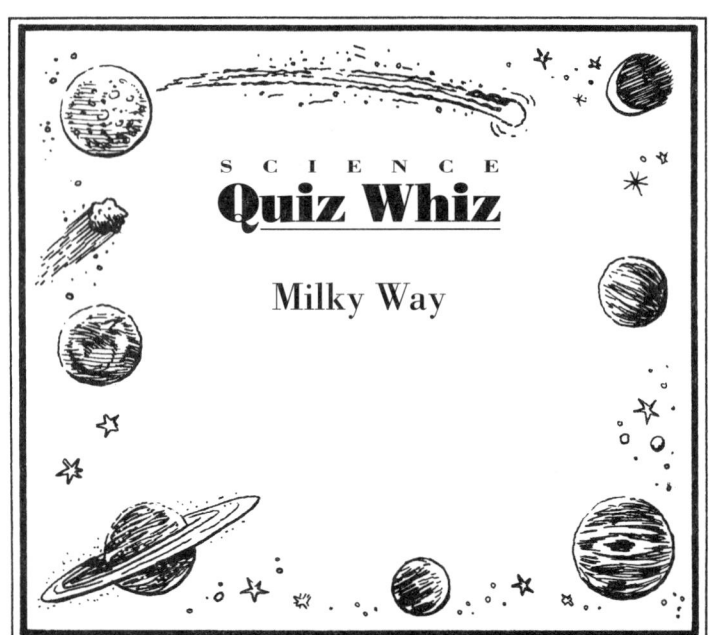

Milky Way

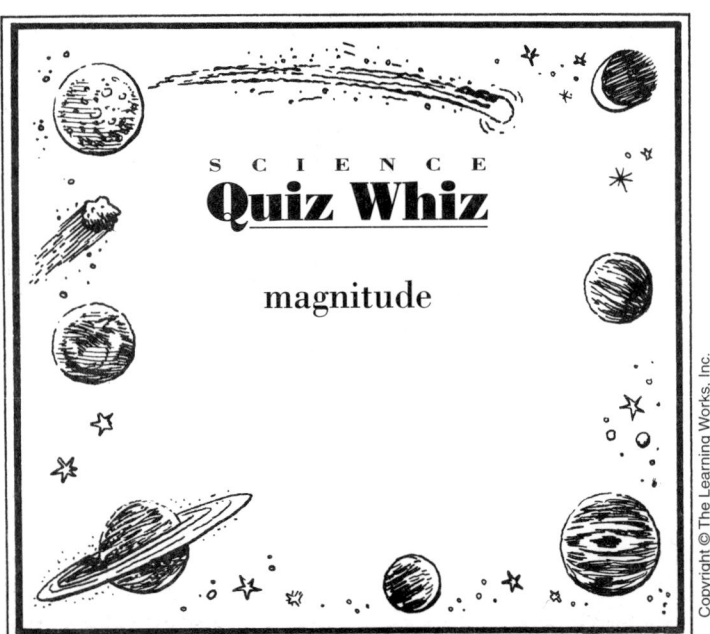

magnitude

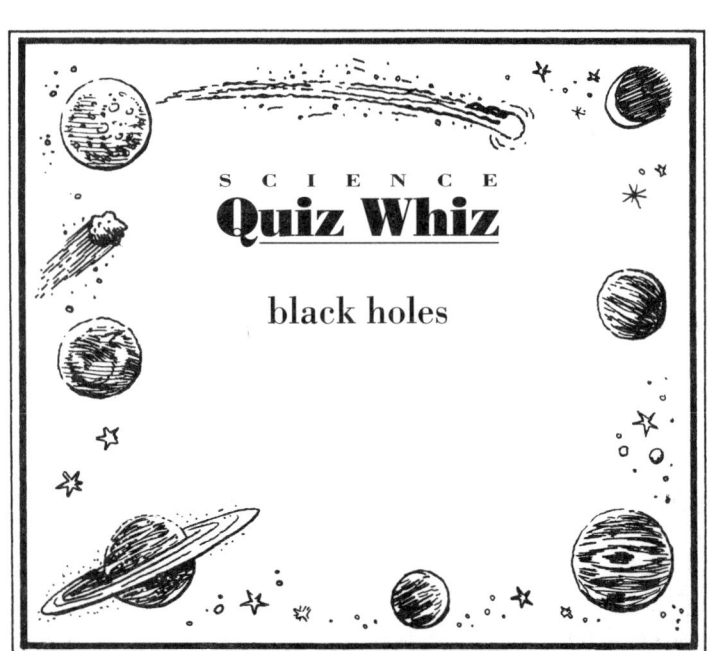

black holes

SCIENCE Quiz Whiz

What are the dark patches on the sun's surface called?

SCIENCE Quiz Whiz

What is the name of the famous comet that passes close to Earth every seventy-six years?

SCIENCE Quiz Whiz

What is the name of the Earth's only natural satellite?

SCIENCE Quiz Whiz

Which planet is the largest in our solar system?

SCIENCE Quiz Whiz

What kind of eclipse occurs when the moon passes between Earth and the sun?

SCIENCE Quiz Whiz

What unit measures the distance that light travels in one year?

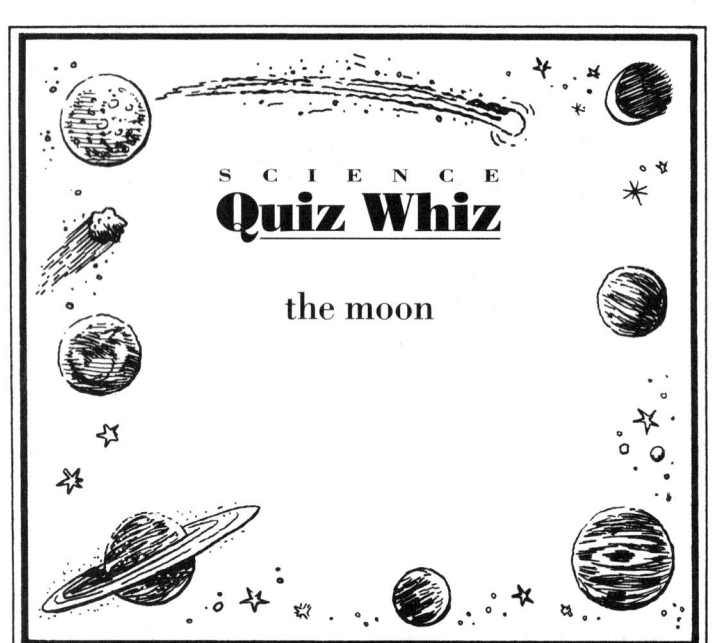

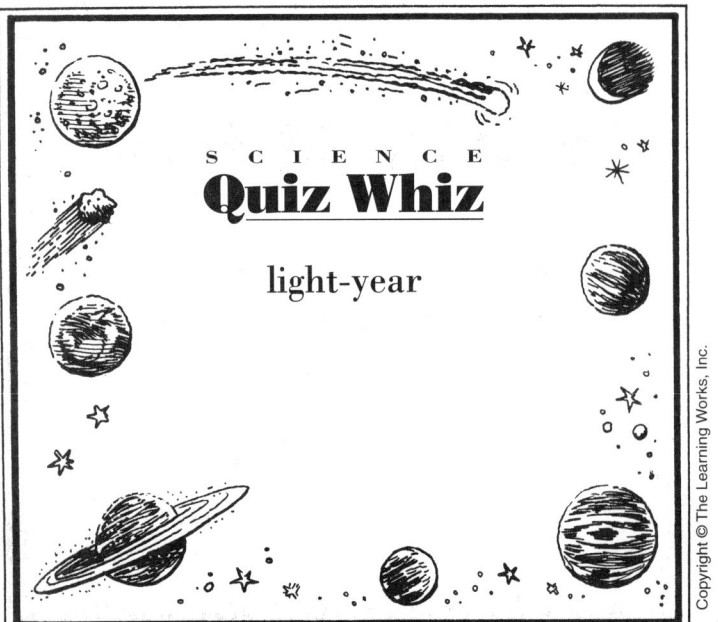

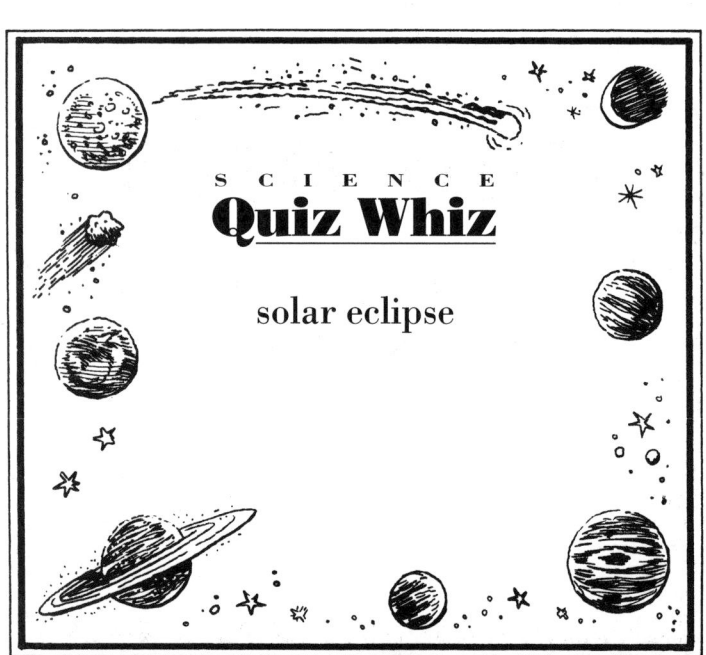

SCIENCE
Quiz Whiz

What is the earth's path around the sun called?

SCIENCE
Quiz Whiz

How many planets are in our solar system?

SCIENCE
Quiz Whiz

Which planet is named for the Roman god of the sea?

SCIENCE
Quiz Whiz

Which is the hottest planet in our solar system?

SCIENCE
Quiz Whiz

How long does it take the earth to travel around the sun?

SCIENCE
Quiz Whiz

What is the name of the force that pulls things toward the center of the earth?

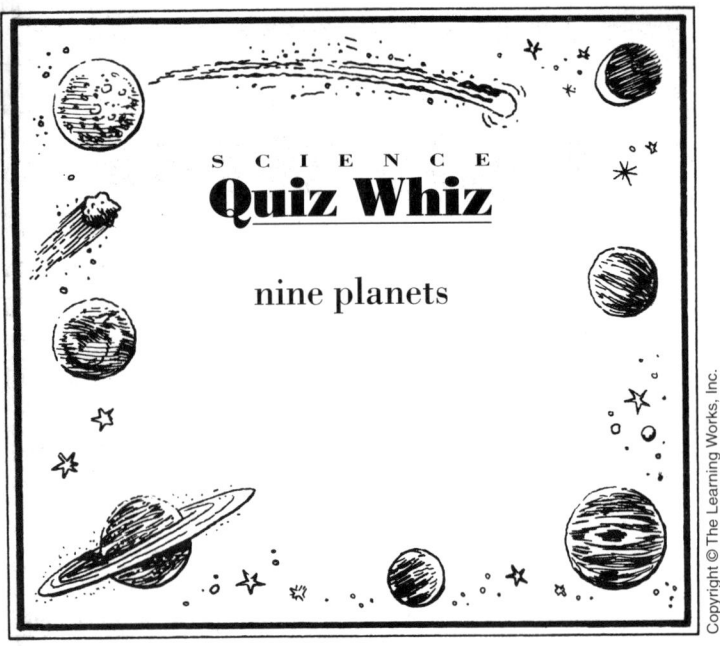

SCIENCE
Quiz Whiz
nine planets

SCIENCE
Quiz Whiz
an orbit

SCIENCE
Quiz Whiz
Mercury

SCIENCE
Quiz Whiz
Neptune

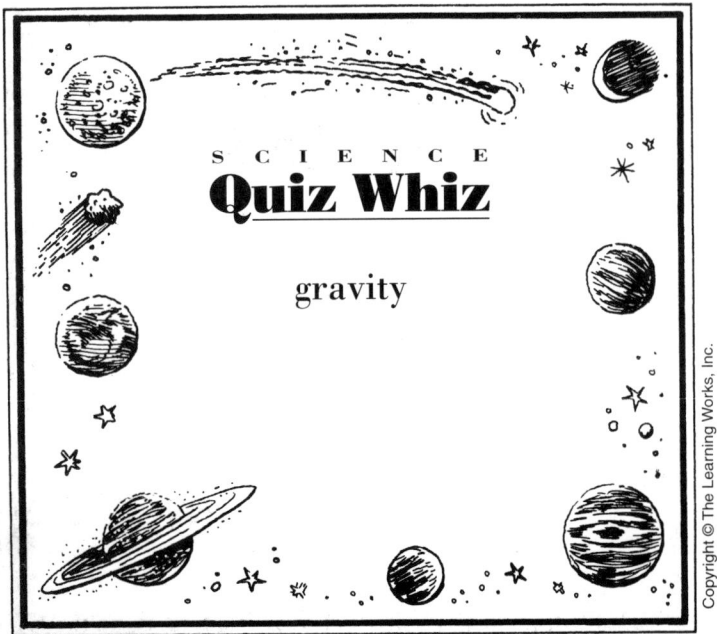

SCIENCE
Quiz Whiz
gravity

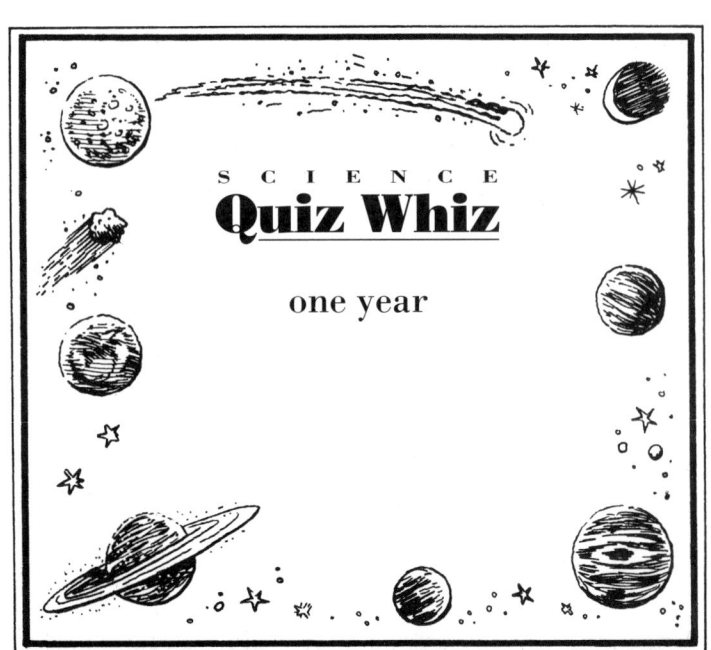

SCIENCE
Quiz Whiz
one year

SCIENCE Quiz Whiz

What nickname is given to the constellation Ursa Major?

SCIENCE Quiz Whiz

What shape do planets trace as they orbit around the sun?

SCIENCE Quiz Whiz

Which red planet is named after the Roman god of war?

SCIENCE Quiz Whiz

Which planet has the longest day?

SCIENCE Quiz Whiz

In searching for extra-terrestrial life, what substance do astronomers look for first: carbon dioxide, water, ammonia, or green cheese?

SCIENCE Quiz Whiz

Which planet is named for the winged messenger in Roman mythology?

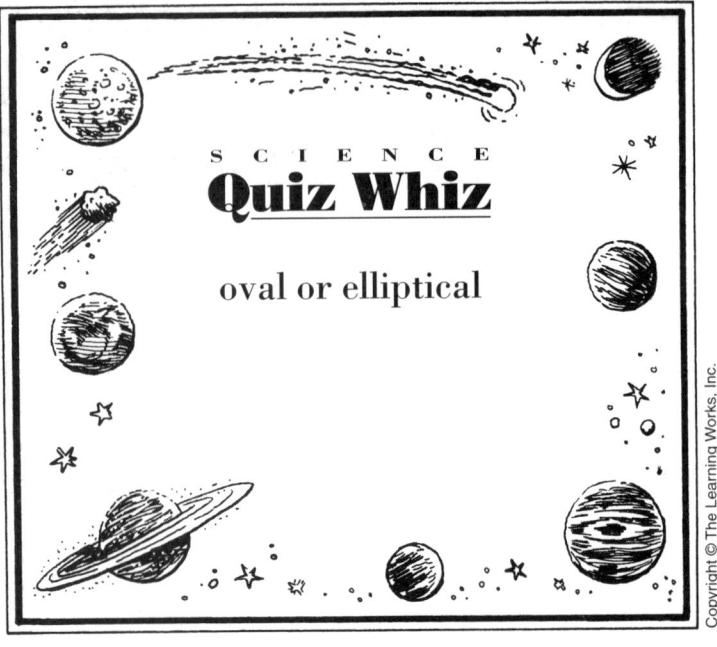

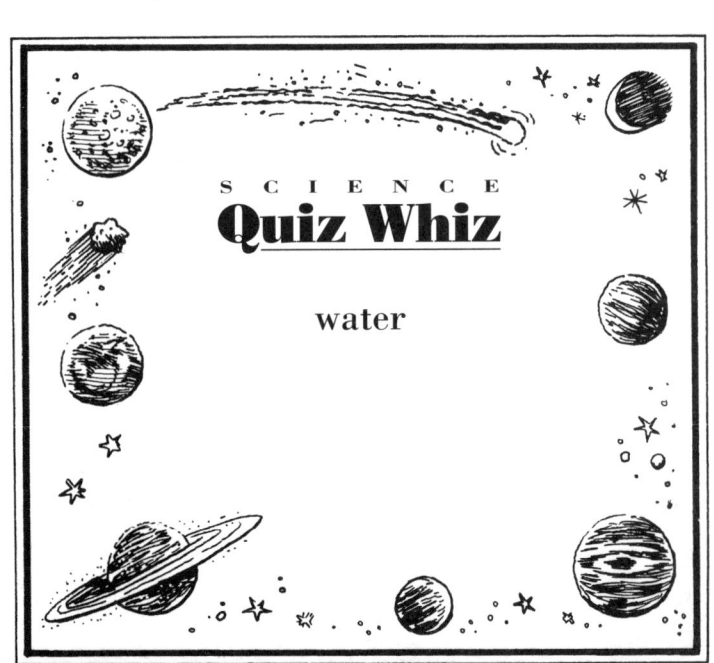

SCIENCE Quiz Whiz

Which planet is named for the ruler of Roman gods?

SCIENCE Quiz Whiz

How long does it take Earth to rotate around on its axis?

SCIENCE Quiz Whiz

How old is our solar system: 4,600 years, 4.6 million years, 4.6 billion years, or 4.6 trillion years?

SCIENCE Quiz Whiz

What is a group of stars called that forms a pattern, such as the Big Dipper?

SCIENCE Quiz Whiz

What is the mixture of gases that surround a planet called?

SCIENCE Quiz Whiz

Which planet has 19 moons and 1,000 rings?

SCIENCE Quiz Whiz

24 hours

SCIENCE Quiz Whiz

Jupiter

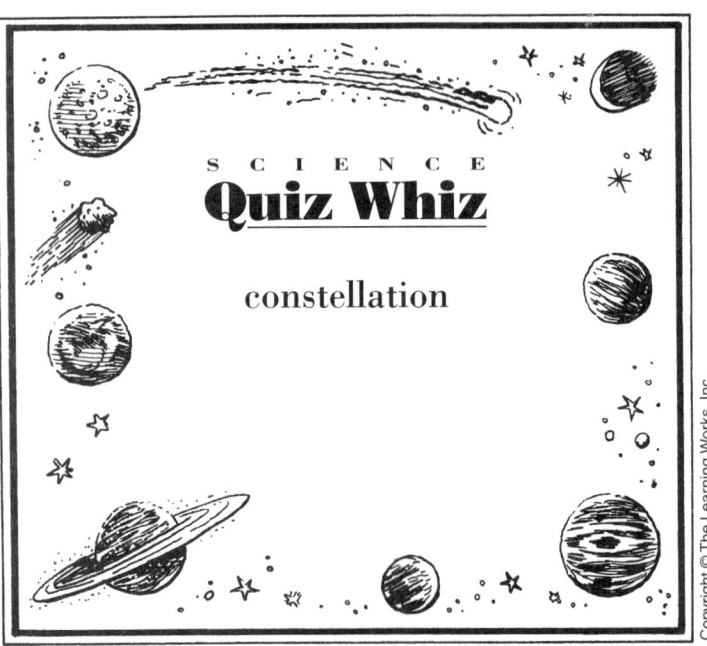

SCIENCE Quiz Whiz

constellation

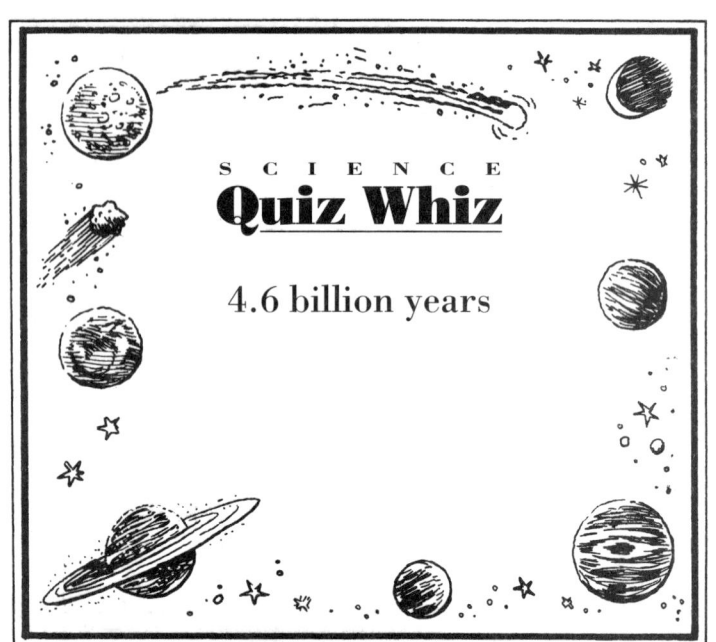

SCIENCE Quiz Whiz

4.6 billion years

SCIENCE Quiz Whiz

Saturn

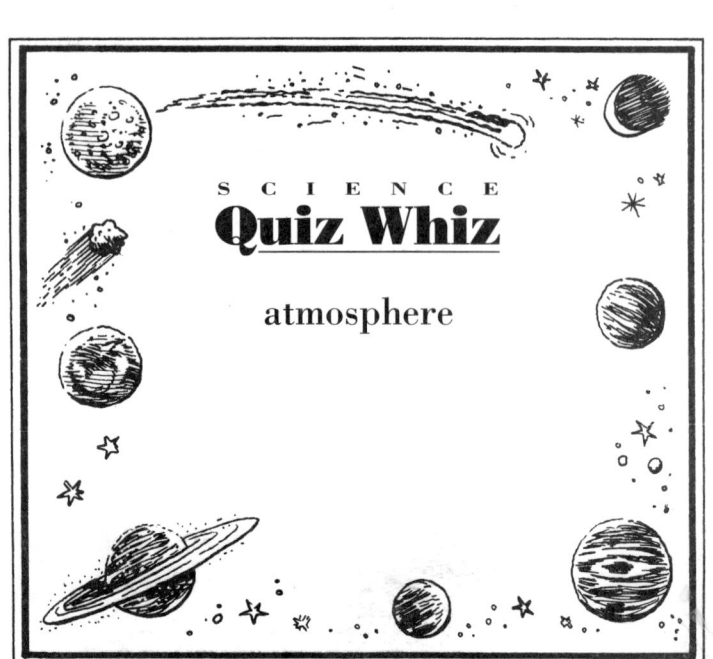

SCIENCE Quiz Whiz

atmosphere